Interfacial Dissipative Phenomena in Tribomechanical Systems

Interfacial Dissipative Phenomena in Tribomechanical Systems

Editor

Antonio Papangelo

MDPI • Basel • Beijing • Wuhan • Barcelona • Belgrade • Manchester • Tokyo • Cluj • Tianjin

Editor
Antonio Papangelo
Mechanics Mathematics and
Management
Politecnico di Bari
Bari
Italy

Editorial Office
MDPI
St. Alban-Anlage 66
4052 Basel, Switzerland

This is a reprint of articles from the Special Issue published online in the open access journal *Lubricants* (ISSN 2075-4442) (available at: www.mdpi.com/journal/lubricants/special_issues/interfacial_dissipative).

For citation purposes, cite each article independently as indicated on the article page online and as indicated below:

LastName, A.A.; LastName, B.B.; LastName, C.C. Article Title. *Journal Name* **Year**, *Volume Number*, Page Range.

ISBN 978-3-0365-2393-4 (Hbk)
ISBN 978-3-0365-2392-7 (PDF)

Contents

About the Editor

Antonio Papangelo

Dr. Antonio Papangelo obtained his Master of Science in Mechanical Engineering from the Polytechnic University of Bari (PoliBa) in 2013. In March 2017 he defended his Ph.D. thesis at PoliBa on the themes of tribology and non-linear dynamics. During his post-doc activity, he has been visiting scientist at the Sandia labs (partner with University of New Mexico), at the VUTC Lab (Imperial College London), and at the Hamburg University of Technology (TUHH). He has been working as research fellow in the Dynamics Group at TUHH for more than two years. He is now Research Fellow at PoliBa while it holds a visiting researcher position at TUHH. Since his Ph.D. studies Antonio has focused on topics related to tribology and non-linear dynamics.

Preface to "Interfacial Dissipative Phenomena in Tribomechanical Systems"

In the last twenty years, tribology and non-linear dynamics have reached major milestones in describing rough contact, friction, damping mechanisms, and dynamical behavior of non-linear systems, which are paving the way for the future engineering technologies. The two fields are largely intertwined as, among the others, contact non-linearities are almost omnipresent in any technical application ranging from the development of NEMS/MEMS to bioengineering, automotive, civil/mechanical industry, and aerospace.

The common thread in both fields is the study of interfaces, particularly of the dissipative phenomena that take place at the interface, providing the source of damping that is exploited to reduce the vibration amplitude of mechanical systems, improving their service life. Despite the great achievements obtained, we are still far from being able to predict the dynamical behavior of mechanical systems involving contact interfaces. Contamination of knowledge between tribology and non-linear dynamics is of outmost importance today to develop strategies to respond promptly to future challenges.

The current Special Issue aimed at bringing together, in the same Issue, contributions from world-leading scientists working in the fields of tribology and non-linear dynamics, with the aim to favor "contamination" of knowledge from the two fields of research.

Antonio Papangelo
Editor

lubricants MDPI

Editorial

Interfacial Dissipative Phenomena in Tribomechanical Systems

Antonio Papangelo [1,2]

1 Department of Mechanics Mathematics and Management, Politecnico di Bari, Via Orabona 4, 70125 Bari, Italy; antonio.papangelo@poliba.it
2 Institute of Structural Dynamics, Am Schwarzenberg-Campus 1, Hamburg University of Technology, 21073 Hamburg, Germany

Keywords: friction; adhesion; dissipation; nonlinear dynamics; contact nonlinearities

check for updates

Citation: Papangelo, A. Interfacial Dissipative Phenomena in Tribomechanical Systems. *Lubricants* **2021**, *9*, 104. https://doi.org/10.3390/lubricants9100104

Received: 8 October 2021
Accepted: 11 October 2021
Published: 15 October 2021

Publisher's Note: MDPI stays neutral with regard to jurisdictional claims in published maps and institutional affiliations.

The last decade has experienced a tremendous development of several technologies that are likely to shape our future. These comprise soft robots, humanoids, autonomous driving, aeronautic, and space technologies. To reduce the development costs, the dynamical behaviour of machines, usually consisting of several jointed components, is often simulated numerically. To ensure a good matching between numerical and experimental results it is of utmost importance to implement reliable and resilient models, which accurately describe the interactions among the different parts of the machine. Indeed, dynamicists are aware that the most difficult part of simulating structure dynamical behaviour is a good understanding of its damping properties. Ensuring a certain service life often coincides with taking care that the vibration amplitude remains below a certain admissible threshold so that the component can confidently fulfil its task along with the machine lifespan. To increase the machine damping properties, dampers are accurately positioned in the structure so that they dissipate as much energy as possible, particularly when the machine is excited close to one of its natural frequencies. Nevertheless, mechanical structures are almost always constituted by several jointed components, hence, a significant share of the energy is dissipated at the contact interfaces. Riveted and bolted joints, seals and bearings all significantly contribute to the damping properties of the assembly and to its dynamical behaviour due to the introduction of nonlinear interaction forces. Nevertheless, after scrutiny of the relevant engineering literature, one finds out that scientific contributions from tribologists rarely take into account the effect of structure dynamic excitation and, on the other hand, dynamicists very often include interfacial characteristics through very simplified empirical models such as the Coulomb friction coefficient and linear contact stiffness.

This Special Issue (SI) was conceived with the intent of encouraging researchers to submit scientific contributions where the interplay between dynamics and tribology is evident. I am pleased to see that this Special Issue has collected seven scientific articles and one review article, all pertaining to the themes of tribology and dynamics.

In Ref. [1] Bonari and Paggi have developed a novel interface finite element procedure to deal with the normal and tangential contact problem of a rigid indenter with an arbitrary profile and a viscoelastic substrate. Through numerical simulations, they have shown that the effect of Coulomb friction and viscoelastic dissipation can be simultaneously investigated, leading to an accurate estimate of the surface tractions, hence, of the energy that is dissipated through the contact interface.

Ref. [2] and [3] have focused on a different source of dissipation, namely that which occurs due to interfacial adhesion. It is well-known from physics that contacting bodies interact through short-range repulsive and long-range adhesive interactions (van der Waals forces). Commonly, adhesion at macroscopic scales is not observed as interfaces are generally randomly rough. By exploiting a multi-asperity representation of a self-affine randomly rough interface with Gaussian distribution of heights, Violano and Afferrante [2] have studied the loading and unloading behaviour of a rough interface, particularly focusing on how the pull-off detachment force and the energy dissipation depend on the surface characteristics. Remarkably, they have shown that dissipation in a loading cycle is proportional to the real contact area, which is in agreement with other analytical [4] and experimental [5] results. Furthermore, they found hysteretic dissipation and pull-off force are much more scattered for surfaces with high fractal dimension. Although roughness, and in particular its height root mean square, is detrimental to macroscopic adhesion, there exist special topographies [6–8] that can be exploited to enhance macroscopic adhesion. Papangelo and Ciavarella in Ref. [3] have used the Boundary Element Method to study the adhesive behaviour of a sphere with superposed an axisymmetric single-wavelength roughness. This geometry has been studied by Guduru and collaborators both analytically [6] and experimentally [7], showing that the pull-off force increases by a factor of about 20 with respect to the smooth sphere. For the Guduru theory to be valid, the contact patch should be simply connected, without inner cracks. In Ref. [3] the contact problem has been solved numerically, also investigating the regions where the theory's hypotheses were not valid. It was shown that the highest pull-off force is reached for a ratio $A/\lambda \approx 10^{-1}$ (A and λ being, respectively, the waviness amplitude and wavelength) as lower values tend to the smooth sphere behaviour with small adhesion enhancement and larger values tend to the appearance of external cracks.

Reference [9] by Zeng and Qi is a good example of how tribological problems are indeed multiphysics, involving several lengths and time scales. Zeng and Qi [9] have studied the process of erosion and corrosion in pipes that transport two-phase flow (gas-solid) encountering a pipe elbow. They have used a multiphysics finite element simulation software to couple turbulent fluid dynamics for the flow, chemical reactions for corrosion and particle dynamics for erosion. Their results show that marked erosion mechanisms are at play on the elbow extrados surface at 40–50°. Furthermore, turbulence influences the concentration of substances, which plays a role in the rate chemical reactions take place (corrosion).

Reference [10] by Genovese et al. is a review on friction and wear test rigs. It is a comprehensive collection of working principles and specifics that are in use to objectively measure friction and wear data. Starting from the first concepts of Leonardo da Vinci, dating back to the 15th century, nowadays different solutions exist, which allow imposing a certain normal load and a relative velocity to two mating interfaces. As noted, only few friction testers allow for temperature control, lubricated or dry contact conditions and permit outdoor measurements to be conducted in real scenarios.

In Ref. [11] Nouira et al. have numerically and experimentally studied the propagation of elastic waves in a mechanical assembly that presents contact interfaces. Indeed, due to the interfacial roughness, the contact stiffness is nonlinear, often represented with a power-law model, which introduces higher-order harmonics into the system dynamical response. The authors have considered the case of two contacting bodies excited by an impulsive force, which was modelled numerically and compared with experimental measurements. Using different contact laws, the authors have shown that accurately modelling the interfacial contact stiffness is crucial to correctly predict the system nonlinear oscillations, proving once more the tight link between tribology and dynamics.

References [12] and [13] dealt with the problem of friction-induced vibrations, which may trigger tedious noise in the proximity of the sliding interface. There are several mechanisms that may trigger friction-induced vibrations; among others, a decaying characteristic of the friction law with sliding velocity. In Ref. [12] Hu and Habib have considered the use of a dynamic vibration absorber (DVA) to suppress undesired vibrations. They have shown that the character of the bifurcation (sub- or supercritical) can be controlled by a proper design of the nonlinear restoring force of the DVA, although the best performance, in terms of reducing the velocity range where stick–slip vibrations occur, is obtained by using a DVA with linear stiffness. Reference [14], by Stender et al., focused on the concept of basin stability in nonlinear systems. Indeed, the classical concept of linear stability is of little help in nonlinear systems where several solutions may coexist in the same parametric region. Linear stability is, in fact, limited to small perturbations around the equilibrium position. On the contrary, given a finite region in the state space where initial conditions are expected to lie, basin stability gives the likelihood for a dynamical system to converge to a certain equilibrium state (static or dynamic). This concept was successfully applied to frictional oscillators that present multiple co-existing stable solutions and proved to be a valuable tool for the analysis of multi-stable systems, particularly when operating conditions are well known.

Finally, the Guest Editor would like to express his sincere gratitude to all authors and reviewers who contributed to this Special Issue and to the editorial staff of Lubricants for their valuable support.

Funding: This research received no external funding.

Institutional Review Board Statement: Not applicable.

Informed Consent Statement: Not applicable.

Conflicts of Interest: The author declares no conflict of interest.

References

1. Bonari, J.; Paggi, M. Viscoelastic Effects during Tangential Contact Analyzed by a Novel Finite Element Approach with Embedded Interface Profiles. *Lubricants* **2020**, *8*, 107. [CrossRef]
2. Violano, G.; Afferrante, L. Roughness-Induced Adhesive Hysteresis in Self-Affine Fractal Surfaces. *Lubricants* **2021**, *9*, 7. [CrossRef]
3. Papangelo, A.; Ciavarella, M. A numerical study on roughness-induced adhesion enhancement in a sphere with an axisymmetric sinusoidal waviness using Lennard–Jones interaction law. *Lubricants* **2020**, *8*, 90. [CrossRef]
4. Deng, W.; Kesari, H. Depth-dependent hysteresis in adhesive elastic contacts at large surface roughness. *Sci. Rep.* **2019**, *9*, 1–12.
5. Dalvi, S.; Gujrati, A.; Khanal, S.R.; Pastewka, L.; Dhinojwala, A.; Jacobs, T.D. Linking energy loss in soft adhesion to surface roughness. *Proc. Natl. Acad. Sci. USA* **2019**, *116*, 25484–25490. [CrossRef] [PubMed]
6. Guduru, P.R. Detachment of a rigid solid from an elastic wavy surface: Theory. *J. Mech. Phys. Solids* **2007**, *55*, 445–472. [CrossRef]
7. Guduru, P.R.; Bull, C. Detachment of a rigid solid from an elastic wavy surface: Experiments. *J. Mech. Phys. Solids* **2007**, *55*, 473–488. [CrossRef]
8. Papangelo, A.; Ciavarella, M. A Maugis–Dugdale cohesive solution for adhesion of a surface with a dimple. *J. R. Soc. Interface* **2017**, *14*, 20160996. [CrossRef] [PubMed]
9. Zeng, Q.; Qi, W. Simulation Analysis of Erosion–Corrosion Behaviors of Elbow under Gas-Solid Two-Phase Flow Conditions. *Lubricants* **2020**, *8*, 92. [CrossRef]
10. Genovese, A.; D'Angelo, G.A.; Sakhnevych, A.; Farroni, F. Review on friction and wear test rigs: An overview on the state of the art in tyre tread friction evaluation. *Lubricants* **2020**, *8*, 91. [CrossRef]
11. Nouira, D.; Tonazzi, D.; Meziane, A.; Baillet, L.; Massi, F. Numerical and experimental analysis of nonlinear vibrational response due to pressure-dependent interface stiffness. *Lubricants* **2020**, *8*, 73. [CrossRef]
12. Hu, J.L.; Habib, G. Friction-Induced Vibration Suppression via the Tuned Mass Damper: Optimal Tuning Strategy. *Lubricants* **2020**, *8*, 100. [CrossRef]
13. Stender, M.; Hoffmann, N.; Papangelo, A. The basin stability of bi-stable friction-excited oscillators. *Lubricants* **2020**, *8*, 105. [CrossRef]
14. Stender, M.; Hoffmann, N. bSTAB: An open-source software for computing the basin stability of multi-stable dynamical systems. *Nonlinear Dyn.* **2021**, 1–18. [CrossRef]

lubricants

MDPI

Article

Viscoelastic Effects during Tangential Contact Analyzed by a Novel Finite Element Approach with Embedded Interface Profiles

Jacopo Bonari *,† and **Marco Paggi** †

IMT School for Advanced Studies Lucca, Piazza San Francesco 19, 55100 Lucca, Italy; marco.paggi@imtlucca.it

* Correspondence: jacopo.bonari@imtlucca.it

† These authors contributed equally to this work.

Received: 20 November 2020; Accepted: 16 December 2020; Published: 19 December 2020

Abstract: A computational approach that is based on interface finite elements with eMbedded Profiles for Joint Roughness (MPJR) is exploited in order to study the viscoelastic contact problems with any complex shape of the indenting profiles. The MPJR finite elements, previously developed for partial slip contact problems, are herein further generalized in order to deal with finite sliding displacements. The approach is applied to a case study concerning a periodic contact problem between a sinusoidal profile and a viscoelastic layer of finite thickness. In particular, the effect of using three different rheological models that are based on Prony series (with one, two, or three arms) to approximate the viscoelastic behaviour of a real polymer is investigated. The method allows for predicting the whole transient regime during the normal contact problem and the subsequent sliding scenario from full stick to full slip, and then up to gross sliding. The effects of the viscoelastic model approximation and of the sliding velocities are carefully investigated. The proposed approach aims at tackling a class of problems that are difficult to address with other methods, which include the possibility of analysing indenters of generic profile, the capability of simulating partial slip and gross slip due to finite slidings, and, finally, the possibility of simultaneously investigating dissipative phenomena, like viscoelastic dissipation and energy losses due to interface friction.

Keywords: viscoelasticity; contact mechanics; finite element method

1. Introduction

A recently developed finite element procedure is herein extended and applied to the analysis of the transient and steady state sliding of a rigid indenter over a deformable material. In accordance with the requirements of current industrial applications, which demand increasingly complex contacting topologies, often down to the micro-scale, together with the analysis of concurrent interface phenomena, like friction and wear, it is shown that the present approach is capable of dealing with arbitrarily complex surfaces and, thanks to the flexibility of the finite element method, to account for any kind of material law.

Indeed, real viscoelastic materials present a time-dependent mechanical response that varies across several orders of magnitude of time and intensity. Therefore, a simple model with a linear Hookean spring in series with a single Newtonian dashpot is far from being representative. For instance, for Ethylene Vynil Acetate (EVA) used as an encapsulating material for photovoltaics, a power-law decay of the Young's modulus with time has been reported [1,2], which can be well-modelled by a fractional viscoelastic model [3–5] as a limit of a Prony series representation with several arms. Its approximation for engineering applications usually requires the use of at least three arms in the Prony series, in order to provide meaningful stress analysis predictions.

In this study, we propose an extension of the variational approach that is based on the interface finite element with eMbedded Profile for Joint Roughness (MPJR) recently proposed in [6,7] for frictionless normal contact problems, and further generalized in [8] in order to simulate frictional partial slip scenarios, to accommodate also finite interface sliding displacements. The methodology, which allows embedding any contact profile as an exact analytical function into an interface finite element, overcomes the cumbersome procedure required by standard finite element methodologies to explicitly discretize the geometry of the boundary exposed to contact. In the MPJR method, the boundary is treated as flat and its actual perturbation from flatness is included as a correction to the normal gap. Since the MPJR method is set to operate within the finite element method (FEM), it presents all the advantages of FEM to solve linear and nonlinear boundary value problems with any arbitrary material constitutive law and structural geometry.

A representative contact problem involving a rigid indenter with harmonic profile acting over a viscoelastic layer of finite depth, perfectly bonded to a rigid substrate, is addressed in order to demonstrate the capabilities of the proposed approach. The loading history will include an applied displacement normal to the contacting interface during a first stage, with a progressive increase in the contact area. Afterwards, the normal displacement is held constant and a horizontal far-field displacement in the sliding direction is applied, in order to simulate the stick-slip transition and then the steady-state sliding regime. Friction is considered along the interface and it is mathematically treated with a regularized Coulomb frictional law. Different sliding velocities, which are relevant for the behaviour of a viscoelastic material, are examined. Numerical simulations provide useful insight into the distribution of the tangential tractions in all of the phases of the sliding process. When considering different Prony series representations with a number of arms varying from one to three, the computational approach allows for quantifying the effect of refining the viscoelastic constitutive model by introducing additional relaxation times.

2. Materials and Methods

2.1. Proposed Solution Scheme for the Contact Problem

In order to investigate the effect of different viscoelastic models along with frictional effects, the contact problem involving a rigid indenter that is characterised by a harmonic profile acting over a layer made of a linear viscoelastic material is addressed. Here, is important to remark that there are no restrictions on the shape of the indenting profile, which can be chosen as an analytical function, or it can be provided as a discrete set of elevations. In the latter case, an external file provided by a profilometer, with a simple two-columns data structure with sampling point coordinate and its elevation, can be used in input. To use such data, one has to keep in mind that the boundary has to be discretized by using MPJR interface finite elements with a uniform spacing dictated by the profilometer resolution, to achieve a one-to-one correspondence between finite element nodes and profilometer sampling points. The assignment of the elevation to each finite element node can be efficiently done only once, just at the beginning of the simulation, by a simple searching algorithm looking for the global coordinate of the finite element node that matches the coordinate stored in the external data file. Subsequently, elevations are efficiently stored in a history variable, in order to avoid multiple reading from external files during the Newton–Raphson iterations and in the next loading steps of the simulation. Further details on the finite element procedure can be found in [6].

2.2. MPJR Formulation

For the solution of the contact problem, the MPJR interface finite element that is exposed in [6,8] is employed. It consists in a 4-nodes, zero-thickness element mutuated by the Cohesive Zone Model (CZM) and used in the context of nonlinear fracture mechanics. The framework is applied to the problem of a rigid body with a complex boundary making contact with another deformable body characterised by a smooth interface. The core of the approach is re-casting the original geometry of the problem into a simpler one, consisting only in the deformable bulk and a single layer of interface elements disposed at its boundary, where contact is supposed to take place, as in Figure 1. The actual shape of the indenter is stored nodal-wise in each interface finite element employed for the interface discretization, and it is used to correct the normal gap function that is computed from the flat–flat configuration, in order to account for the exact geometry. This requires a preliminary step, which consists in mapping the indenter profile elevation in correspondence to the right node of the boundary. If the profile has analytic expression, this can be done right at the finite element level exploiting the global coordinates, otherwise the elevation field can be stored in a proper history variable and every entry associated with the correct node. It has to be remarked that, in spite of the present formulation being $2D$, the proposed framework can be extended to $3D$ problems, provided that, for example, a 8-nodes interface element is used to discretize a surface, instead of a profile, equipped with a suitable friction law.

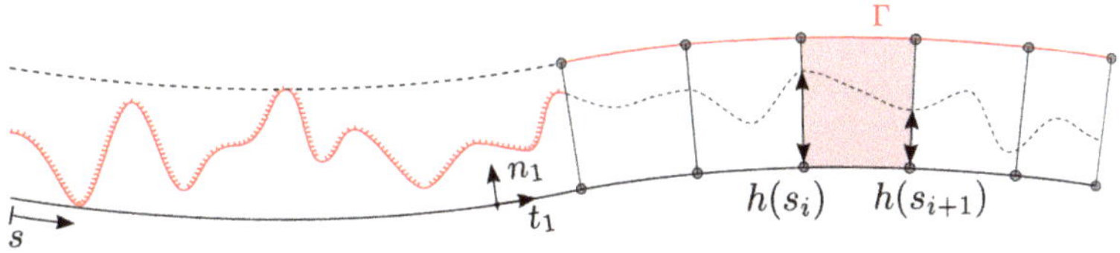

Figure 1. Profile discretization and equivalent interface definition. The interface element Γ is defined with the lower two nodes that belong to the deformable bulk, and the others placed at a given offset normal to the lower boundary. An abscissa s can be defined along the boundary to map the indenter's elevation field, which is stored inside the element and it is used to correct the normal gap.

Figure 2 shows the kinematics of the element. A vector of unknown nodal displacements $\mathbf{u} = [u_1, v_1, \dots, u_4, v_4]^T$ is introduced for the evaluation of the tangential and normal gaps, collected in the vector $\mathbf{g} = [g_x, g_z]^T$, which reads:

$$\mathbf{g} = \mathbf{QNLu}, \tag{1}$$

where $\mathbf{L}$ is a linear operator for computing the relative displacements across the interface, $\mathbf{N}$ is the shape functions matrix, and $\mathbf{Q}$ is a rotation matrix for transforming displacements from the global to the local reference frame of the element defined by the unit vectors n and t. The original geometry can be restored with a suitable correction of the normal gap, in the form:

$$\mathbf{g}^* = \begin{bmatrix} g_x \\ g_z + h(\xi, t) \end{bmatrix}, \tag{2}$$

where $h(\xi, t)$ maps the profile's shape and position in time. With respect to the formulation that is presented in [8], here the profile shape has been made time-dependent, in order to also account for finite sliding of the rigid indenter. For example, in the case of a flat interface, the result of the indenter sliding with a given constant velocity v_0 can be achieved by setting $h(\xi, t) = h(\xi - v_0 t)$.

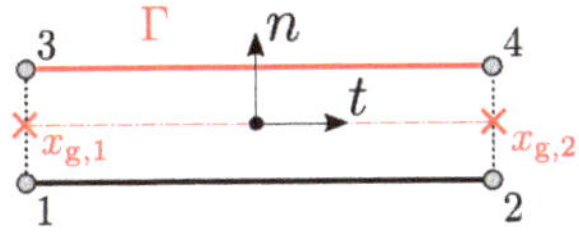

Figure 2. Four-nodes, zero-thickness eMbedded Profiles for Joint Roughness (MPJR) interface finite element.

The current value of t is stored at the interface element level while using a time history variable, and it is updated every time step. A standard penalty approach is used in order to enforce the normal contact constraint, leading to

$$p_z = \begin{cases} \alpha g_z^*, \text{ if } g_z^* < 0, \\ 0, \text{ if } g_z^* \geq 0, \end{cases} \tag{3}$$

where α is the penalty parameter.

To deal with coupled frictional problems, a regularized Coulomb friction law [9] is used to set the interface constitutive equation in the tangential direction:

$$q_x = f p_z \tanh\left(\frac{\dot{g}_x}{\dot{\varepsilon}}\right), \tag{4}$$

where q_x is the tangential traction and $\dot{g}_x$ is the sliding velocity, as given by the difference between the velocity of the indenter and the horizontal velocity of the corresponding node. Finally, $\dot{\varepsilon}$ is a parameter governing the slope of the regularised friction law.

The contribution of a single interface finite element to the variational formulation of the bulk material is expressed by its variation in terms of density of energy content, integrated over the domain Γ_e that denotes the element itself:

$$\delta\Pi_e = \int_{\Gamma_e} \delta\mathbf{g}^*(\mathbf{v})^T \mathbf{p}(\mathbf{u}, \dot{\mathbf{u}})\, d\Gamma_e, \tag{5}$$

where $\mathbf{v}$ is the virtual displacement field, and the vector $\mathbf{p}$ collects the normal and tangential tractions. As a final step, the variation can be expanded and the integral set to zero, leaving the expression of the nonlinear residual vector, which reads:

$$\mathbf{R}_e(\mathbf{u}, \dot{\mathbf{u}}) = \int_{\Gamma_e} \mathbf{L}^T \mathbf{N}^T \mathbf{Q}^T \mathbf{p}(\mathbf{u}, \dot{\mathbf{u}})\, d\Gamma_e = \mathbf{0}. \tag{6}$$

Because of the nonlinearity of $\mathbf{R}_e$, the Newton–Raphson iterative method has been applied, together with a backward Euler method for time integration.

2.3. Rheological Model

Three different Prony series models with a number of arms increasing from one to three are examined in order to assess the effect of viscoelasticity modelling on the overall contact mechanical response. The general equation for the shear relaxation modulus reads:

$$\frac{G(t)}{G^\infty} = \mu_0 + \sum_{n=1}^{3} \mu_n \exp\left(-\frac{t}{\tau_n}\right), \tag{7}$$

where G^∞ is the instantaneous shear modulus (evaluated at $t = 0$), μ_n are the relaxation coefficients, and τ_n are the corresponding relaxation times. Equation (7) has been tuned to fit the experimental values of EVA [4]. The model parameters for 1, 2 and 3 arms are collected in Table 1.

Table 1. Rheological parameters for Ethylene Vynil Acetate (EVA), where n is the number of Prony series' arms.

n [–]	G^{∞} [Pa]	μ_0 [–]	μ_n [–]	τ_n [s]
1	568.498	0.421	0.579	0.817
2	674.606	0.306	0.398	0.212
			0.296	2.458
3	749.386	0.254	0.310	0.102
			0.226	0.545
			0.210	4.104

The identification of the above parameters has been carried out through a regression over the experimental data that were acquired in the time range $t = 10^{[-1,\ldots,+1]}$ s. The following approach has been pursued in order to attain a high degree of accuracy. Firstly, trial relaxation times have been set and a preliminary linear regression has been performed involving G^{∞} and μ_i only. The objective function to be minimised reads:

$$\Pi(\mathbf{x}) = \sum_{k=1}^{N} \left(\mathbf{g}_k \cdot \mathbf{x} - G_k\right)^2, \tag{8}$$

where, for the three arms model, $\mathbf{g}_k = \left[1, e^{(-t_k/\tau_1)}, \ldots, e^{(-t_k/\tau_3)}\right]$, G_k is the value of the objective function at the sampling point and N is the number of samplings. The global minimiser $\mathbf{x}^* = \arg\min_x \Pi(\mathbf{x})$ is evaluated and the values of the constants μ_i and G^{∞} are obtained according to:

$$G^{\infty} \begin{bmatrix} \mu_0 \\ \mu_1 \\ \mu_2 \\ \mu_3 \end{bmatrix} = \mathbf{x}^*, \tag{9}$$

together with the condition $\sum_i \mu_i = 1$, related to the shear modulus at $t = 0$. The obtained coefficients, together with their respective relaxation times, have been used in order to define a vector of guess values $\mathbf{x}_0$ for a second nonlinear regression, in which the relaxation times were also included in the optimisation vector $\mathbf{x}$. The problem has been solved iteratively, updating the starting vector $\mathbf{x}_0$ every cycle using the results that were obtained in the previous. Convergence is achieved within 5 iterations, when considering a relative error that is given by $(\mathbf{x}^* - \mathbf{x}_0)/\mathbf{x}_0$ and a tolerance $\varepsilon = 10^{-15}$. This procedure has also been repeated in the same way for the 1 and 2 arms models.

Once the parameters are identified, the Young's relaxation modulus $E(t)$ can be obtained from $G(t)$, and the behaviour of the three models can be investigated in time and frequency domains. The analysis in the frequency domain can be performed by defining a complex modulus $\hat{E}(\omega)$, obtained via a Fourier transform of $E(t)$, which can be expressed as:

$$\frac{\hat{E}(\omega)}{E^{\infty}} = \mu_0 + \sum_{i=1}^{n} \mu_i \frac{\tau_i^2 \omega^2}{1 + \tau_i^2 \omega^2} + \imath \sum_{i=1}^{n} \mu_i \frac{\tau_i \omega}{1 + \tau_i^2 \omega^2}. \tag{10}$$

In the expression above, $\imath$ denotes the imaginary unit and the index k defines the number of arms being considered. It can be easily noticed that, for the single arm model, the maximum viscoelastic effect manifests in correspondence to the critical excitation frequency $\omega^{\star} = \sqrt{\mu_0}/\tau_1$.

Figure 3a shows the plot of $E(t)$. Figure 3b,c show the values of the loss modulus and the storage modulus, which were obtained as the imaginary part $\Im\hat{E}(\omega)$ and the real part $\Re\hat{E}(\omega)$ of the complex modulus $\hat{E}(\omega)$, respectively. Finally, Figure 3d shows the loss tangent, given as the ratio of the imaginary part over the real part. As a comparison, the same quantities are also plotted for the relaxation modulus obtained for a model that is based on fractional calculus, which reads:

$$E_{\mathrm{f}}(t) = \frac{E_{\mathrm{f},\alpha} t^{-\alpha}}{\Gamma(1-\alpha)}. \tag{11}$$

In Equation (11), $E_{\mathrm{f},\alpha} = 814.7\,\mathrm{Pa\,s}^{\alpha}$ and $\alpha = 0.226$ have been chosen in order to fit the experimental data in [4], being $\Gamma(\cdot)$ the gamma function.

The simulation of the power-law viscoelastic response seen in the experiments, which is well approximated by the fractional calculus model, is progressively improved by increasing the number of terms in the Prony series representation. It has to be remarked that, since the Fourier transform of a power-law is a power-law itself, both loss and storage modulus in the frequency domain are represented, on a logarithmic scale, as straight lines.

Their trend can be satisfactory modelled with Prony series only for a narrow band of the whole spectrum, based on the relaxation time(s) employed. Therefore, the relaxation times entering Prony series have to be regarded as design parameters, to be chosen based on the loading history experienced by the viscoelastic material, rather than material parameters. With the values that were chosen here, an accurate estimation of the material response can be expected, at most, over two orders of magnitude, centred on a frequency of 1 Hz.

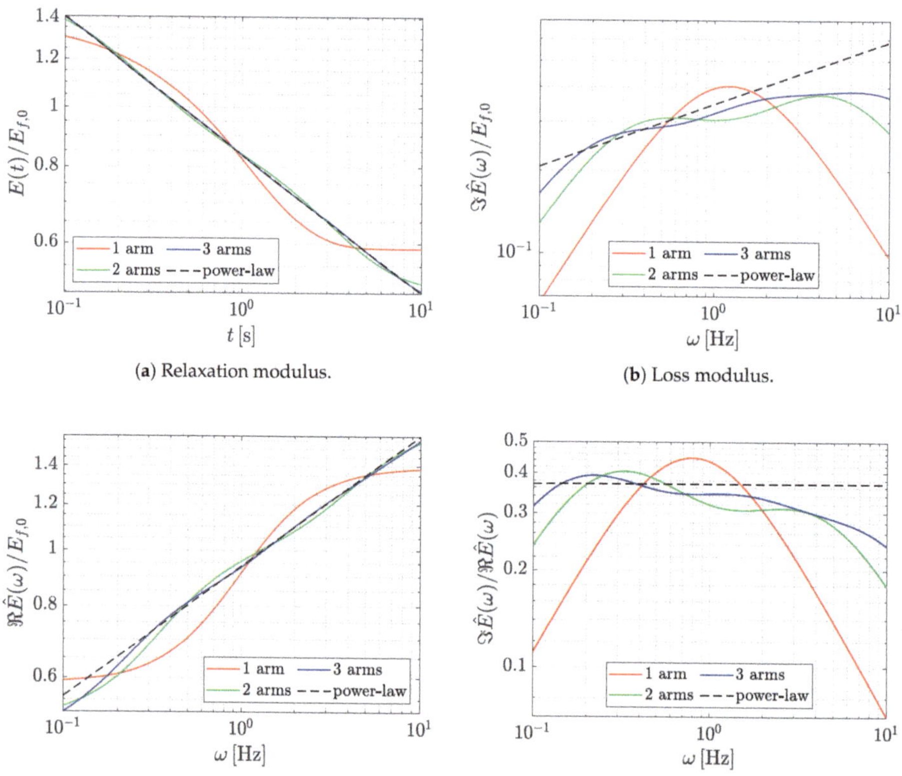

(**a**) Relaxation modulus. (**b**) Loss modulus. (**c**) Storage modulus. (**d**) Tangent modulus.

Figure 3. Relaxation modulus in time and frequency domain.

2.4. Problem Set Up

We focus our attention onto a displacement controlled problem under plane strain assumptions in order to highlight the capability of the proposed approach. In the first stage, a displacement linearly increasing with time is applied along the direction normal to the finite layer, up to a given final value of $\Delta_{z,0} = 2g_0$, reached at time $t = t_0$, which is then held constant. At this point, a tangential displacement with a constant horizontal velocity is applied to the indenter, which starts sliding. The indenter profile is analytically expressed by:

$$\frac{h(x,t)}{g_0} = 1 - \cos\left[\frac{2\pi}{\lambda_0}(x - vt)\right] \tag{12}$$

While the velocity of the application of normal load is the same for all the simulations, and assumed to be quasi-static, for what concerns the horizontal load different sliding velocities have been considered in the range $v_i = 10^{(i-10)/3}[\mathrm{m/s}]$, $i = [1, \ldots, 10]$, with their numerical value being summarised in Table 2.

Table 2. Range of horizontal velocities employed.

v [m/s]
1.000×10^{-03}
2.154×10^{-03}
4.642×10^{-03}
1.000×10^{-02}
2.154×10^{-02}
4.642×10^{-02}
1.000×10^{-01}
2.154×10^{-01}
4.642×10^{-01}
$1.000 \times 10^{+00}$

A regularized Coulomb frictional law [8] is considered, with $f = 0.2$ being the friction coefficient. Figure 4 lists the remaining geometric parameters that describe the problem set, together with the rheological model that is employed for modelling viscoelasticity, which has already been thoroughly discussed in Section 2.3: three different simulations are performed, each of them characterised by one, two, or three terms of a Prony series used for modelling a linear viscoelastic material. The model geometry and applied velocities are the same in all of the cases considered. Finally, periodic boundary conditions have been introduced in correspondence of the two vertical sides of the domain, in order to simulate a semi-indefinite contact in the horizontal direction. The simulations have been performed using the Finite Element Analysis Program FEAP [10], where the MPJR formulation has been implemented as a user element routine. The validation of the proposed computational method is provided in Appendix A.

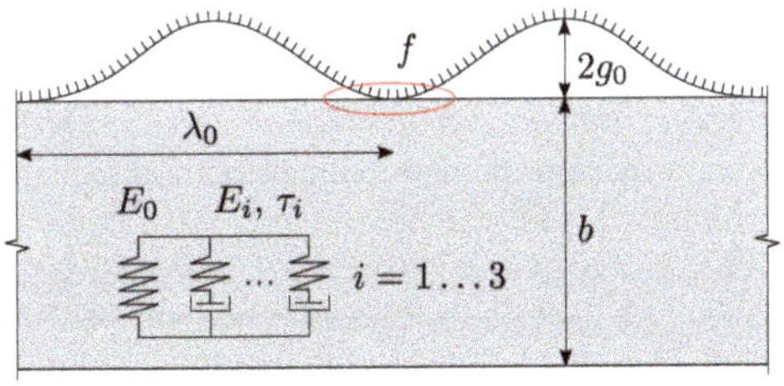

Figure 4. Sketch of the model, $b = 1$, $\lambda_0 = b$, $g_0 = 5 \times 10^{-4}\lambda_0$.

3. Results

3.1. Bulk Stresses

Figure 5 shows the results of FEM simulations for the boundary value problem shown in Figure 4. They refer to the single arm model, but, from a qualitative point of view, the considerations that are going to be drawn below for the bulk stresses also apply to the other two models herein considered.

Figure 5a,b display the stresses developing in the bulk at the end of the normal loading stage, and they display three distinct areas with high stresses where the harmonic profile comes into contact. Because of the presence of friction and, since coupling effect are fully included, an anti-symmetric distribution of τ_{xz} arises, even in the pure normal loading stage, see Figure 5b. The following two figures represent the same quantities at a subsequent load stage, where the normal imposed displacement has reached its maximum, and the indenter slides at constant velocity. Figure 5c,d show the stresses during the next stage of sliding, corresponding to a lateral shift of the harmonic profile of about half of its wavelength. The advantage of the finite element method is evident from the possibility to consider any finite-size problem geometry and boundary conditions, like, in this case, the output automatically including not only contact tractions, but also bulk stresses.

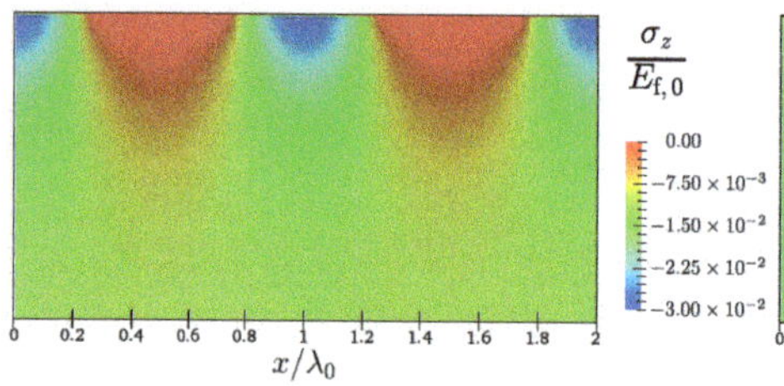

(**a**) σ_z, normal loading stage.

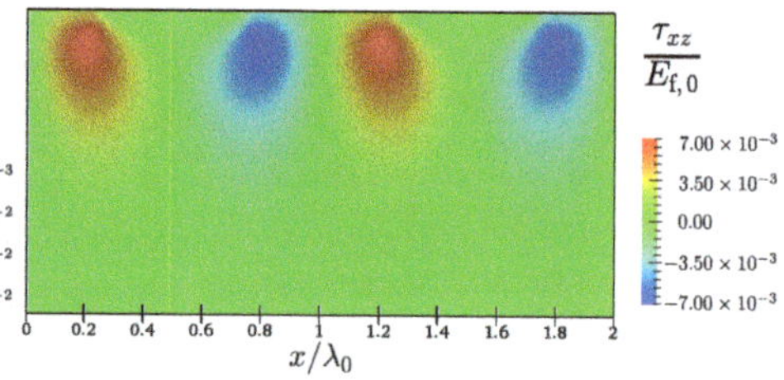

(**b**) τ_{xz}, normal loading stage.

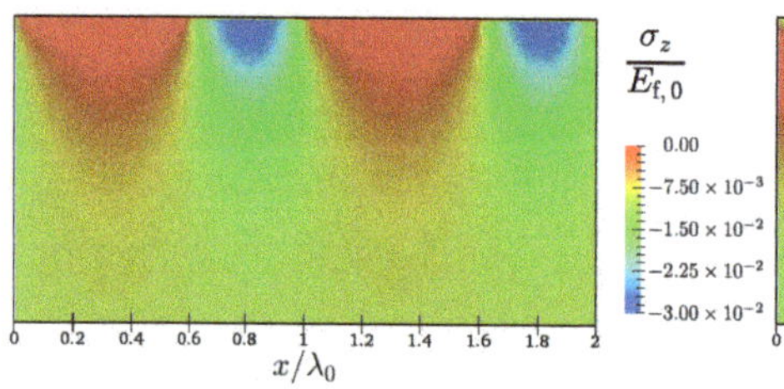

(**c**) σ_z, tangential loading stage.

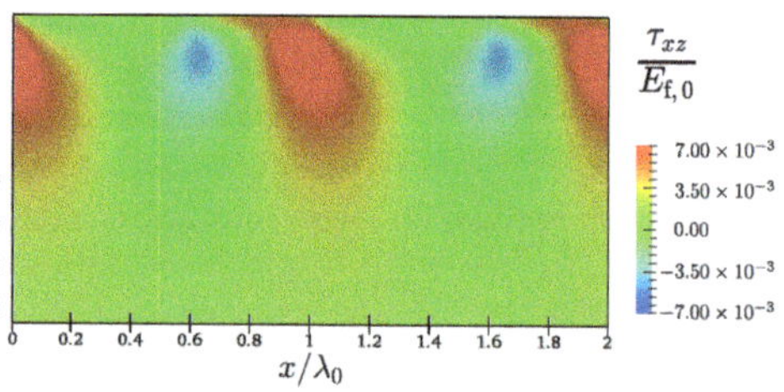

(**d**) τ_{xz}, tangential loading stage.

Figure 5. Model predictions: bulk stresses during the normal approach, (**a**,**b**), and during full sliding, (**c**,**d**), all scaled by a reference elastic modulus $E_{f,0} = 8.147 \times 10^2$ Pa.

3.2. Interface Tractions

Figure 6 highlights the evolution of contact tractions in time for the single arm model and selected stages of the contact simulation. The curves in Figure 6a correspond to the purely normal loading sequence, where normal contact tractions progressively increase along with the value of the applied normal displacement, which linearly rises from zero up to the final value of $2g_0$. Black curves denote the symmetric distribution of normal contact tractions $p_z(x)$ divided by $E_{f,0}$, while red curves represent the anti-symmetric distribution of tangential contact tractions $q_x(x)$, scaled by $fE_{f,0}$. Points along the interface, where $\|q_x(x)\|/(fE_{f,0})$ equals $p_z(x)/E_{f,0}$, are in a state of slip, while, when the inequality $\|q_x(x)\|/(fE_{f,0}) < p_z(x)/E_{f,0}$ holds, then there is a state of stick.

Figure 6b refers to the next stage of the contact problem when, keeping the normal displacement constant, a far-field displacement linearly increasing with time is applied in the tangential direction. While, for the given rheological model, the results that are shown in Figure 6a are evaluated in a condition of zero tangential velocity, Figure 6b–d are referred to $v = 2.154 \times 10^{-2}$ m/s. This specific value has been chosen amid the other entries of Table 2, because it is in the middle of the range, determining the highest viscoelastic effects, and it is also low enough for analysing the transition from *stick/slip* to *full sliding*, Figure 6b. Here, tangential traction distributions change their shape from the classical anti-symmetric form towards a state of increasing slip, which terminates in the full slip condition. The transition from *stick-slip* to *full slip* is strongly affected by the velocity of the horizontal displacement: the faster the slip, the more abrupt such a transition.

Figure 6c refers to the situation of sliding after full slip (*gross sliding*) and, in particular, it shows the evolution over time of the normal contact tractions. We see a transition from the symmetric contact traction distribution along the whole interface at the onset of full slip, as shown in black, towards other distributions in different scales of grey shifted along the interface to the right, as long as the tangential displacement increases. A certain degree of relaxation is observed after the onset of full slip. As the sliding proceeds in time, virgin material is perturbed, and a recovery in stiffness takes place.

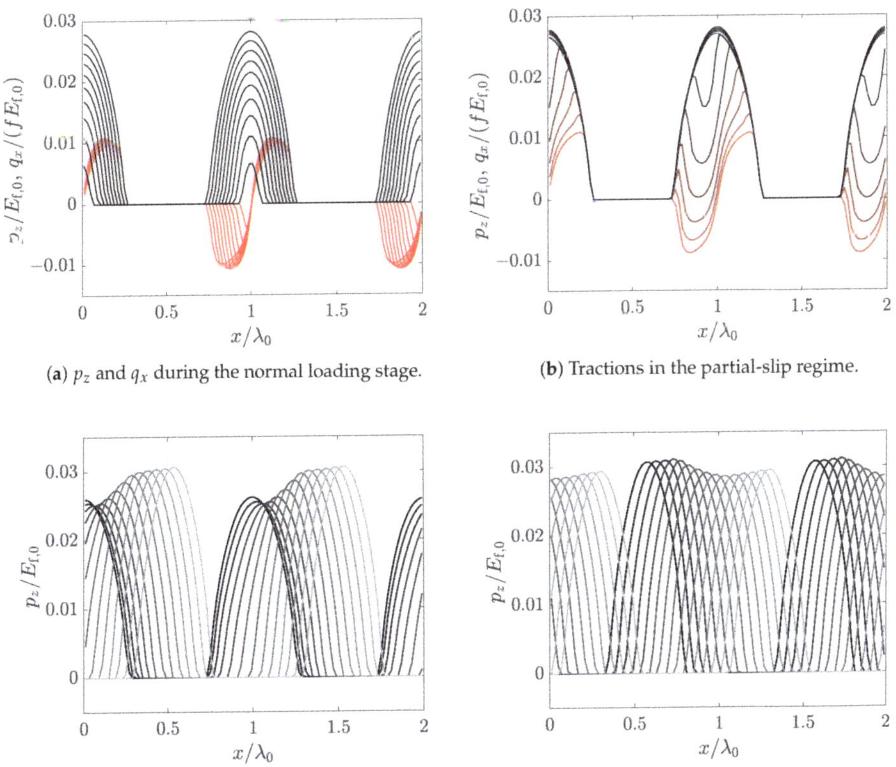

(**a**) p_z and q_x during the normal loading stage.

(**b**) Tractions in the partial-slip regime.

(**c**) Normal traction during full sliding.

(**d**) Normal traction during interaction with an already stressed portion of the interface.

Figure 6. Selected distributions of normal and tangential contact tractions during the different stages of loading.

Finally, Figure 6d captures the first overlapping of a new contact zone with a previously loaded portion of the interface. Here, the role of the relaxation time is important, since viscoelastic effects do alter the solution that corresponds to a linear elastic material that has no memory effects.

The resultant tangential force Q_x, integral of tangential contact tractions along the interface, is plotted vs. time in Figure 7a–c for the three viscoelastic models investigated herein. In each subfigure, different curves correspond to different far-field horizontal displacement velocities. Darker curves correspond to slower velocities.

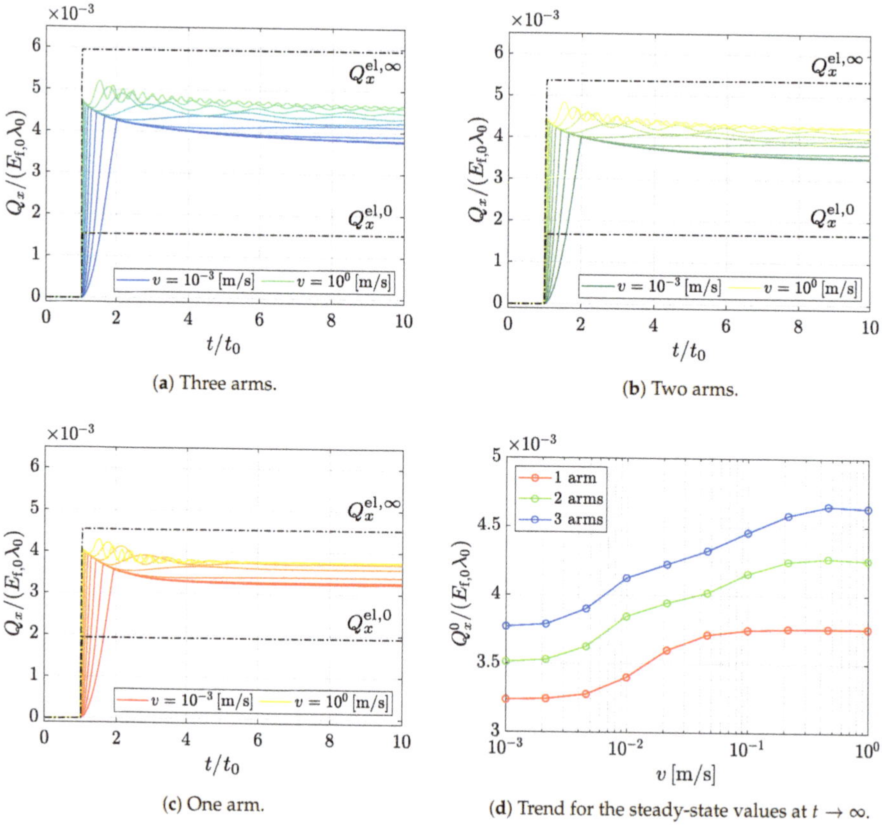

(**a**) Three arms. (**b**) Two arms. (**c**) One arm. (**d**) Trend for the steady-state values at $t \to \infty$.

Figure 7. Time evolution of the resultant tangential force Q_x for different rheological models.

In all of the cases, for $t/t_0 \leq 1$, tangential tractions are vanishing, since, in that stage, the imposed displacement is only acting in the normal direction. Therefore, tangential contact tractions are due to frictional coupling effects and their sum over the whole contact zones is vanishing by definition, since they correspond to self-equilibrated distributions. For $t/t_0 > 1$, the indenter starts sliding and we assist to a transition from *stick-slip* to *full slip* with an oscillatory behaviour when the contact profile enters in contact with unrelaxed material portions. When the velocity is low, no rate effects are evident, and the mechanical response is smooth. On the other hand, by increasing the applied velocity, the importance of viscoelasticity increases and oscillating responses do appear.

The integral of tangential tractions related to two linear elastic models that are characterised by short and long term modulus are also plotted in Figure 7; for comparison, see black dash-dotted lines. The elastic moduli are evaluated as:

$$E^{\mathrm{el},\infty} = \lim_{t \to 0} E(t) = E^{\infty} \qquad E^{\mathrm{el},0} = \lim_{t \to \infty} E(t) = E^{\infty}(1 - \sum_{i=1}^{n} \mu_i) \tag{13}$$

The curves $Q_x^{\mathrm{el},\infty}$ and $Q_x^{\mathrm{el},0}$ are evaluated under the assumption of linear elasticity, neglecting the dynamic effects. For this reason, they lead to constant values as soon as the horizontal far-field displacement is applied, without any oscillation. The only factor that plays a role is the velocity, which governs the transition from *stick/slip* to full sliding. In the figures, only the curves that correspond to the highest value of v are plotted. In all three models, the instantaneous (higher) and long term (lower) curves are extreme bounds to the values that are related to viscoelastic simulations, with a gap increasing from the single arm to the three arms model, consistent with their respective stiffness.

The steady-state solution strongly depends on the rheological properties of the material, as shown in Figure 7d. In general, for the present case study, the higher the number of arms, the higher the total tangential force. In all cases, the highest velocity determines the highest value of the steady state Q_x^0. This is in accordance with the fact that, in a condition of *gross slip*, $Q_x = fN_z$, and for high velocities, the material is excited in its high frequency region, thus resulting in a vertical response that is governed by the higher *glassy* Young's modulus. The increased stiffness leads to higher N_z^0 and, in turn, higher Q_x^0 values.

4. Conclusions

In this study, a novel finite element procedure has been proposed, which allows for investigating transient and steady state sliding of a rigid indenter over a viscoelastic continuum. In particular, the representative problem of an indenter with harmonic profile sliding over a viscoelastic layer of finite depth has been analysed, employing different sliding velocities together with three different rheological models, which are characterised by Prony series with one, two, and three arms, respectively. A regularised version of the classic Coulomb friction law has been employed for the evaluation of the interface tangential tractions.

Numerical results pinpoint a strong dependence of the mechanical response in terms of steady-state forces N_z and Q_x on both the velocity and rheological model employed, obtaining increasing forces for higher velocities and more relaxation terms that are involved in the rheological approximation.

It is worth mentioning that the proposed methodology appears to be suitable for the investigation of a class of problems for which a solution could be difficult to be found while using other techniques. The proposed approach is capable of overcoming the limitations of other solution schemes thanks to the capability of FEM of solving linear and nonlinear boundary value problems with arbitrary material constitutive laws and geometries. Moreover, the use of the recently developed interface finite element [6–8] has further advantages. First of all, the possibility of taking into account arbitrary shapes for the indenting profile as analytical functions that are embedded into the interface element. The ability of simulating partial slip scenarios involving finite sliding of the indenter should also be mentioned.

Moreover, as a key advantage when compared to other models that are available in the literature that neglect the effect of Coulomb friction, focusing on viscoelastic dissipation only, here viscoelastic effects and frictional effects can be simultaneously investigated, since they are inherently coupled in the formulation. Neglecting interface tangential tractions, together with their related coupling affecting the distribution of normal tractions, could be reasonable when incompressibility conditions are approached. On the other hand, several evidences can be found that, as the Young's modulus of a viscoelastic material changes with time, so does the Poisson's ratio. Because the latter quantity governs the coupling between normal and tangential tractions, a fully coupled model is worth study for fine precision engineering applications. As a final remark, the proposed interface finite element has the further advantage of being easily extended for taking thermal effects into account. These could be relevant not only for the analysis of temperature transfer across the interface, but also to simulate frictional heat generation, thus leading to a thermodynamically accurate model that is capable of investigating a wide class of realistic viscoelastic dissipative phenomena.

Author Contributions: Conceptualization, J.B. and M.P.; methodology, J.B. and M.P.; software, J.B.; validation, J.B.; formal analysis, J.B. and M.P.; investigation, J.B. and M.P.; resources, M.P.; data curation, J.B.; writing—original draft preparation, J.B.; writing—review and editing, M.P.; visualization, J.B.; supervision, M.P.; project administration, M.P.; funding acquisition, M.P. All authors have read and agreed to the published version of the manuscript.

Funding: This research received funding from the Italian Ministry of University and Research through the Research Project of National Interest (PRIN 2017) "XFAST-SIMS: Extra fast and accurate simulation of complex structural systems" (grant no. D68D19001260001).

Conflicts of Interest: The authors declare no conflict of interest.

Appendix A. Model Validation

The proposed framework has been tested against a Hertz indentation problem for validation. The solution of the FEM simulation is compared with the analytical solution of the equivalent half-plane 2D contact problem, in terms of the integral of the interface normal and tangential tractions P_z and Q_x, respectively, given the same imposed displacements history. A parabolic profile has been used as a first order approximation of a circular rigid cylinder with unitary radius R_i. The profile makes contact on the flat side of a linear elastic semi-disk with plane strain Young's modulus $E^* = 814.7\,\mathrm{Pa}$ and radius $R_\mathrm{d} = 5R_\mathrm{i}$, which simulates a half-plane. The load history includes two far-field displacements, imposed to the rigid profile. First, a normal displacement is applied, starting from zero and linearly increasing to a maximum value $\Delta_{z,0}/R_\mathrm{i} = 1 \times 10^{-3}$, reached at time t_0, see the black line in Figure A1. The normal displacement is then held constant, and a harmonic tangential displacement is applied, which increases up to a maximum $f\Delta_{z,0}$, being $f = 0.2$ the coefficient of friction, and then makes a complete cycle, see the red line in Figure A1. Such a maximum value of horizontal displacement is chosen to cause the incipient sliding of the cylinder, and this is indeed what happens if the response of the system in terms of frictional reaction forces is analysed.

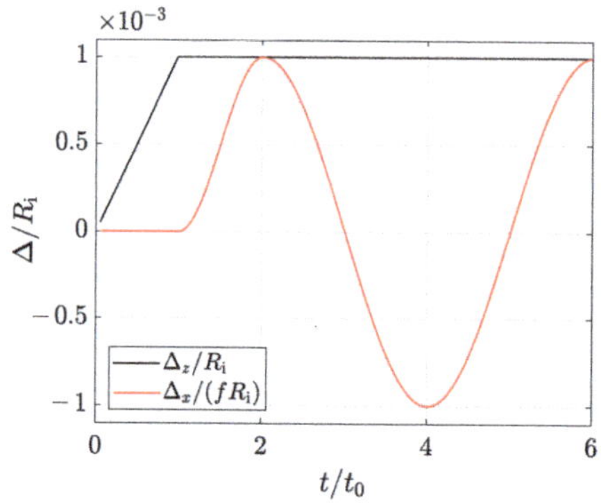

Figure A1. Imposed displacements.

Appendix A.1. Evaluation of Normal Reaction Forces

For plane contact problems, displacements can only be evaluated to within an arbitrary constant or, equivalently, in reference to a datum point. For the 2*D* Hertz problem, the boundary displacements normal to the interface can be evaluated as ([11], pp. 20–24):

$$w(x) = \begin{cases} -\frac{2P_z}{\pi E^*}\left[\left(\frac{x}{a}\right)^2 + c_0\right], \text{ if } x \leq a, \\ -\frac{2P_z}{\pi E^*}\left[\log|\psi(x)| + \frac{1}{2\psi(x)^2} + \frac{1}{2} + c_0\right], \text{ if } x \geq a, \end{cases} \tag{A1}$$

where c_0 is the arbitrary constant, and:

$$\psi(x) = \frac{x}{a} + \sqrt{\left(\frac{x}{a}\right)^2 - 1}. \tag{A2}$$

An additional equilibrium equation relates the value of the load with the extension of the contact semi-strip a:

$$a = \sqrt{\frac{4P_z R_i}{\pi E^*}}. \tag{A3}$$

If the datum is set in correspondence of the point of the boundary $x = R_d$, the relation between the imposed displacement and the resultant vertical load has the form:

$$w(0) - w(R_d) = \Delta_z = \frac{2P_z}{\pi E^*}\left[\log \psi(R_d) + \frac{1}{2\psi(R_d)^2} + \frac{1}{2}\right], \tag{A4}$$

where $w(0)$ is evaluated in coincidence of the point of first contact, coincident with the centre of the semi-disk. As a final step, the inversion of Equation (A4) for a given value of Δ_z gives the desired P_z. The comparison with numerical results is shown in Figure A2, where diamond markers representing the FEM prediction show a very good accordance with the corresponding solid black line, that represents the analytical results.

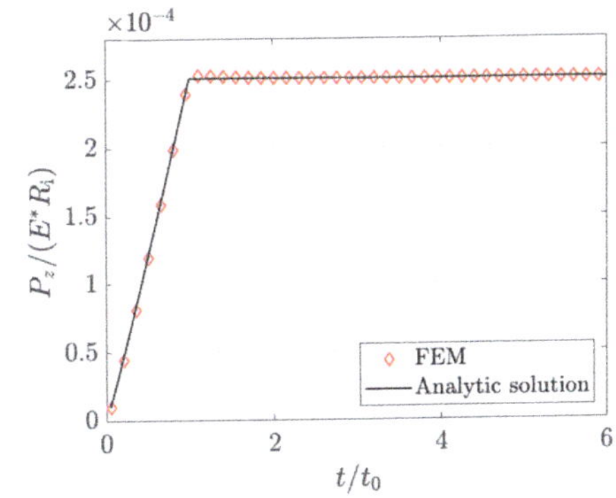

Figure A2. Resulting integrals of surface normal tractions.

Appendix A.2. Evaluation of Tangential Reaction Forces

Finding Q_x for a given displacement still requires the evaluation of the applied displacement history with respect to a reference value, still set in correspondence of $x = R_d$. Since a closed form solution is not available for the tangential tractions, an extended version of the Jäger-Ciavarella theorem that accounts for variable normal and tangential loads have been used for evaluating the analytical solution of the problem, according to the algorithm presented in [12]. If a load path is defined in terms of Δ_z and Δ_x, then, according to the theorem, the tangential problem can be reduced to the normal one, since an increment in tangential forces can be evaluated as the difference between the actual vertical force and the vertical force related to a smaller imposed vertical displacement, multiplied by the coefficient of friction:

$$Q_x = f\left[P_z(\Delta_z) - P_z(\Delta_z^*)\right]. \tag{A5}$$

The value of Δ_z^* is a function of Δ_x. For a constant normal load and an increasing tangential load, it can be evaluated as:

$$\Delta_z^* = \Delta_z - \frac{\Delta_x}{f}. \tag{A6}$$

For general loading scenarios, the principle can be extended and the correct value of Δ_z^* evaluated in terms of an equivalent path that respects both the equilibrium and the friction law. Results are shown in Figure A3,

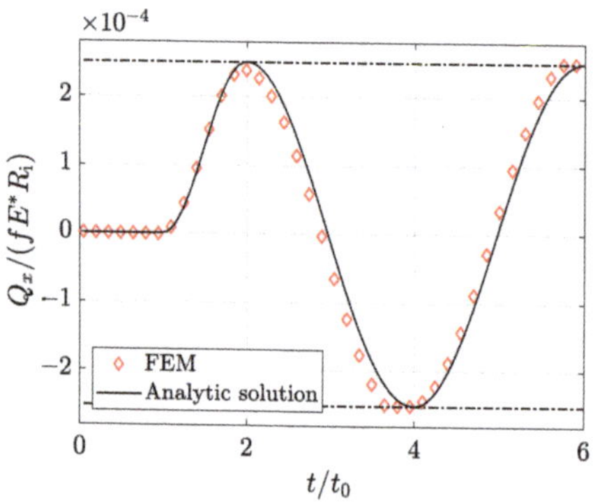

Figure A3. Resulting integrals of surface tangential tractions.

Where good accordance is found between the analytical solution given by the solid black line and the numerical prediction, depicted by the red diamond markers. In the same figure, the limit of *gross slip* for forward and backward sliding is shown as well by means of positive and negative valued horizontal black dash-dotted lines, respectively. These values represent the upper and lower threshold for the values of Q_x, and this condition is approached in correspondence of the related maximal tangential imposed displacement, cfr. Figure A1.

As a final remark, the differences between the numerical and the analytical results, for both normal and tangential forces, are due to the effect of coupling between normal and tangential tractions, which is not taken into account by the analytical approach. Moreover, another source of the small difference lies in the treatment of the friction law: FEM exploits a regularised Coulomb friction law, while the analytical approach exploits the classical one, where the stick-slip transition is abrupt.

References

1. Eitner, U. Thermomechanics of Photovoltaic Modules. Ph.D. Thesis, Zentrum für Ingenieurwissenschaften der Martin-Luther-Universität Halle-Wittemberg, Halle, Germany, 2011.
2. Eitner, U.; Kajari-schröder, S.; Köntges, M.; Brendel, R. Non-linear mechanical properties of ethylene-vynil acetate (EVA) and its relevance to thermomechanics photovoltaic modules. In Proceedings of the 25th European Conference on Photovoltaic Solar Energy, Valencia, Spain, 6–10 September 2010; pp. 4366–4368.
3. Paggi, M.; Kajari-Schröder, S.; Eitner, U. Thermomechanical deformations in photovoltaic laminates. *J. Strain Anal. Eng. Des.* **2011**, *46*, 772–782. [CrossRef]
4. Paggi, M.; Sapora, A. An Accurate Thermoviscoelastic Rheological Model for Ethylene Vinyl Acetate Based on Fractional Calculus. *Int. J. Photoenergy* **2015**, *2015*. [CrossRef]
5. Gagliardi, M.; Lenarda, P.; Paggi, M. A reaction-diffusion formulation to simulate EVA polymer degradation in environmental and accelerated ageing conditions. *Sol. Energy Mater. Sol. Cells* **2017**, *164*, 93–106. [CrossRef]

6. Paggi, M.; Reinoso, J. A variational approach with embedded roughness for adhesive contact problems. *Mech. Adv. Mater. Struct.* **2020**, *27*, 1731–1747. [CrossRef]
7. Paggi, M.; Bemporad, A.; Reinoso, J. Computational Methods for Contact Problems with Roughness. In *Modeling and Simulation of Tribological Problems in Technology*; Paggi, M., Hills, D., Eds.; Springer International Publishing: Cham, Switzerland, 2020; pp. 131–178. [CrossRef]
8. Bonari, J.; Paggi, M.; Reinoso, J. A framework for the analysis of fully coupled normal and tangential contact problems with complex interfaces. **2020**, submitted.
9. Wriggers, P. *Computational Contact Mechanics*; Springer: Berlin/Heidelberg, Germany, 2006. [CrossRef]
10. Zienkiewicz, O.; Taylor, R. *The Finite Element Method for Solid and Structural Mechanics*; Butterworth-Heinemann: Oxford, UK, 2000; Volume 2.
11. Hills, D.; Nowell, D. *Mechanics of Fretting Fatigue*; Springer-Science + Business Media: Berlin/Heidelberg, Germany, 2020; pp. 20–24.
12. Jäger, J. A New Principle in Contact Mechanics. *J. Tribol.* **1998**, *120*, 677–684. [CrossRef]

Publisher's Note: MDPI stays neutral with regard to jurisdictional claims in published maps and institutional affiliations.

lubricants

MDPI

Article

Roughness-Induced Adhesive Hysteresis in Self-Affine Fractal Surfaces

Guido Violano * and **Luciano Afferrante**

Department of Mechanics, Mathematics and Management, Polytechnic University of Bari, Via E. Orabona, 4, 70125 Bari, Italy; luciano.afferrante@poliba.it
* Correspondence: guido.violano@poliba.it

Abstract: It is known that in the presence of surface roughness, adhesion can lead to distinct paths of loading and unloading for the area–load and penetration–load relationships, thus causing hysteretic loss. Here, we investigate the effects that the surface roughness parameters have on such adhesive hysteresis loss. We focus on the frictionless normal contact between soft elastic bodies and, for this reason, we model adhesion according to Johnson, Kendall, and Roberts (JKR) theory. Hysteretic energy loss is found to increase linearly with the true area of contact, while the detachment force is negligibly influenced by the maximum applied load reached at the end of the loading phase. Moreover, for the micrometric roughness amplitude h_{rms} considered in the present work, adhesion hysteresis is found to be affected by the shorter wavelengths of roughness. Specifically, hysteresis losses decrease with increasing fractal dimension and cut-off frequency of the roughness spectrum. However, we stress that a different behavior could occur in other ranges of roughness amplitude.

Keywords: adhesion hysteresis; rough surfaces; JKR theory

Citation: Violano, G.; Afferrante, L. Roughness-Induced Adhesive Hysteresis in Self-Affine Fractal Surfaces. *Lubricants* **2021**, *9*, 7. https://doi.org/10.3390/lubricants9010007

Received: 1 December 2020
Accepted: 4 January 2021
Published: 7 January 2021

1. Introduction

The hysteretic dissipation is given by the difference between the work needed to bring two bodies into contact and that required to detach them. Its origin may be related to various phenomena occurring at the contact interface. The main causes of hysteresis are viscoelasticity [1,2], plasticity [3], adhesive elastic instabilities at jump-in and jump-out of contact [4], and surface roughness [5]. In particular, all natural and artificial surfaces are rough at some scale. Therefore, hysteretic losses may affect several technological applications. For example, biomedical devices [6] and structural adhesives [7] must safely adhere to surfaces during their application, but they should be easy to remove for reuse. Moreover, a recent challenge in soft robotics is to create climbing robots with reversible adhesion skills [8].

In contact experiments on soft matter, velocity-dependent dissipations are usually measured during detachment as a consequence of bulk viscoelasticity [9]. In a recent work [5], Dalvi et al. carried out loading–unloading contact experiments between smooth silicone hemispheres and rough nanodiamond substrates. Their experiments were performed at very low velocities (60 nm/s) both for the approach and detachment. Such choice allows the avoidance of velocity-dependent dissipations. However, great adhesion hysteresis was still observed due to the roughness-induced increase in the true contact area. This effect is expected to occur in compliant materials with small root mean square (rms) roughness amplitude ($h_{\mathrm{rms}} \simeq 1$ nm) [1], when they are bring in complete contact. Moving from the assumption of full-contact conditions, Dalvi et al. applied Persson and Tosatti (PT) adhesion theory [10] for predicting the magnitude of adhesion hysteresis. They found that the hysteretic dissipation increases almost linearly with the true contact area A, and it is equal to the product between A and the intrinsic surface energy $\Delta\gamma$, which depends on the interfacial adhesive properties of contacting bodies.

In [11], it is experimentally shown that h_{rms} can both increase and decrease the effective adhesive surface energy with respect to the smooth case. Moreover, numerical simulations of continuum adhesive contacts [12] have shown that there is an optimal h_{rms} that leads to a maximization of the hysteretic loss and pull-off force. Such value of h_{rms} is found when the contact region turns from being simply connected to being multiply connected. Similarly, in [13] it is shown that the effective surface energy $\Delta\gamma_{eff}$ reaches a maximum for a certain h_{rms} that, for vanishing applied pressure, is quite close to the value above which the effective contact area A becomes smaller than the nominal one A_0. Moreover, the enhancement in the adhesion for small h_{rms} is much larger for $H < 0.5$, where H is the Hurst exponent, as the roughness-induced increase in the surface area is smaller when $H > 0.5$.

For RMS roughness amplitudes of the order of few microns, the true area of contact is expected to be predominantly multiply connected. In such case, partial contact conditions occur and surface roughness leads to a reduction in the true area of contact. This in turn destroys adhesion. However, Kesari et al. [11] found that adhesion hysteresis can also be measured for relative large h_{rms}. Inspired by the experimental findings in [11], Deng and Kesari (DK) [14] developed an analytical model for estimating hysteresis losses in the adhesive elastic contacts under the assumption of large roughness. DK's model captures the increase of adhesion hysteresis with the penetration, which is usually called depth-dependent hysteresis. Moreover, in this case, a linear increase of the hysteretic dissipation with the area of contact is observed.

Carbone et al. [15] developed a numerical code based on a Boundary Element Method (BEM) for predicting loading–unloading hysteresis loops in the adhesive elastic contact of fractal self-affine 1D rough profiles. Their simulations were conducted under partial contact conditions, with A/A_0 ranging from 0.25 up to 0.5. Due to adhesion hysteresis, two distinct paths were obtained for loading and unloading curves of the area vs. load relation. In particular, they found two sources of energy dissipation, one occurring at small scales and the second one at large scales.

In [16,17], similar multiasperity models have been developed to estimate adhesion hysteresis. They moved from the pioneering Greenwood and Williamson (GW) model, in which roughness is described by a distribution of identical spherical asperities. Adhesion is then implemented according to the classical theory of Johnson, Kendall, and Roberts (JKR) [18]. In a loading–unloading cycle, each asperity exhibits a hysteretic dissipation, which is due to jump-in and jump-off contact instabilities. The total adhesion hysteresis is returned by the contribution of each asperity. However, such models are based on a simplistic description of the surface roughness and do not take into account the elastic coupling between contact regions.

In this work, we propose an investigation of the adhesive elastic contact of rough surfaces, described by self-affine fractal geometries, with an advanced multiasperity model taking into account lateral interactions of asperities according to the authors of [19,20] and adhesion according to JKR theory. Moreover, the model takes also account of the jump-in and jump-off contact instabilities occurring on each asperity.

2. Problem Statement

2.1. *Adhesion Hysteresis of Smooth Elastic Spheres*

JKR theory [18] is commonly used to predict adhesion of soft materials. In the case of spherical contact, the fundamental equations of JKR theory give the applied load F and penetration δ as a function of the contact radius a

$$F = \frac{4}{3}\frac{E^* a^3}{R} - \sqrt{8\pi E^* \Delta\gamma a^3} \quad (1)$$

$$\delta = \frac{a^2}{R} - \sqrt{\frac{2\pi a \Delta\gamma}{E^*}}, \quad (2)$$

being R the radius of curvature, E^* the composite elastic modulus of the contacting bodies and $\Delta\gamma$ the interface adhesion energy.

Under controlled displacement conditions, JKR model predicts a jump into contact at $\delta = 0$. However, Wu [21], investigating the jump-in instability occurring in atomic force microscopy measurements, found that jump-in instability is reached at a critical gap δ_{in}. Such effect is due to van der Waals interactions acting between approaching bodies. Wu proposed an empirical formula for the jump-in distance (valid for $\mu \geq 2$),

$$\delta_{\text{in}} = \left(1 - 2.641\mu^{3/7}\right)\epsilon \tag{3}$$

where $\mu = \left(\Delta\gamma^2 R/E^{*2}\right)^{1/3}/\epsilon$ is the so-called Tabor parameter [22] and ϵ is the range of attractive forces.

The above equation can be used to modify JKR theory to consider the jump-in critical distance (see in [23]).

Moreover, under displacement controlled conditions and during retraction, JKR theory predicts a jump-off instability at a critical penetration

$$\delta_{\text{off}} = -\left(\frac{27\pi^2\Delta\gamma^2 R}{64E^{*2}}\right)^{1/3}. \tag{4}$$

Figure 1 shows the loading-unloading cycle predicted by JKR theory. The yellow area represents the energy loss due to jumping instabilities. For smooth contact, the energy loss is independent on the maximum penetration (or, equivalently, applied force).

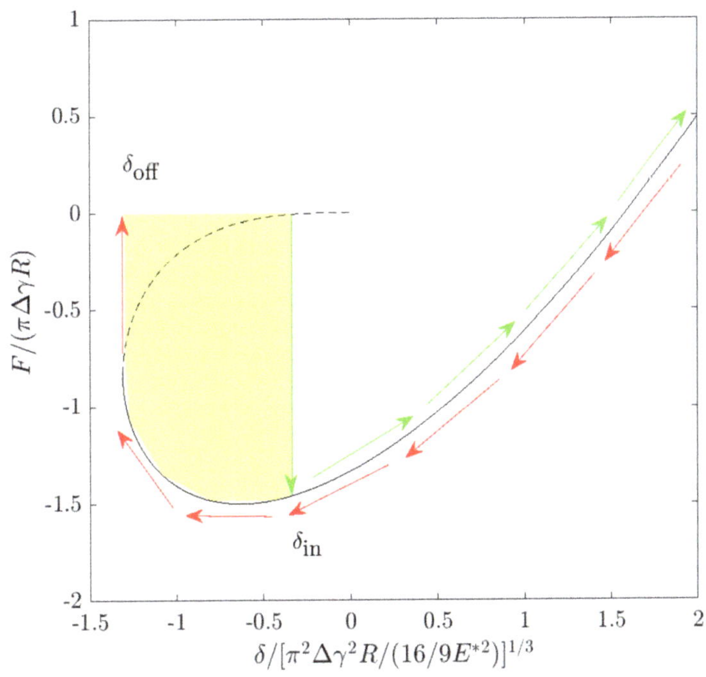

Figure 1. The load–penetration curve predicted by Johnson, Kendall, and Roberts (JKR) theory. The loading (unloading) path is denoted by green (red) arrows. Positive penetration values correspond to indentation and compressive force. Jump into contact occurs at a penetration δ_{in}. The unloading path overlaps the loading one, but jump out of contact occurs at a critical penetration δ_{off}. The yellow area denotes the hysteretic energy loss.

2.2. Adhesion Hysteresis of Rough Elastic Surfaces

In the case of rough surfaces, multiple unstable jumps occur at the location of each asperity in a loading–unloading cycle. To take account of this phenomenon, Equations (3) and (4) are implemented in a multiasperity model which in turn takes into account the elastic coupling due to asperities lateral interactions.

Let us consider a rigid rough surface approaching an elastic half-space (Figure 2). According to JKR formalism, the normal displacement w_i of the elastic half-space at the location of the asperity i is

$$w_i = \frac{a_i^2}{R_i} - \sqrt{\frac{2\pi a_i \Delta\gamma}{E^*}} + \hat{w}_i \tag{5}$$

where $\hat{w}_i$ is the displacement due to the elastic interaction between the asperities in contact and is given by [24]

$$\hat{w}_i = \sum_{j=1, j\neq i}^{n_{ac}} \frac{a_j^2}{\pi R_j}\left(\sqrt{\frac{r_{ij}^2}{a_j^2} - 1} + \left(2 - \frac{r_{ij}^2}{a_j^2}\right)\arcsin\left(\frac{a_j}{r_{ij}}\right)\right) \tag{6}$$

$$-\frac{1}{\pi a_j E^*}\sqrt{8\pi a_j^3 E^* \Delta\gamma}\arcsin\left(\frac{a_j}{r_{ij}}\right) \quad \text{for } r_{ij} > a_j \tag{7}$$

$$\hat{w}_i = \frac{a_j^2}{R_j} - \sqrt{\frac{2\pi a_j \Delta\gamma}{E^*}} \quad \text{for } r_{ij} \leq a_j \tag{8}$$

where n_{ac} is the number of contact spots and r_{ij} is the distance between the asperities i and j.

When the rough surface approaches the half-space, a new contact is formed when the gap between an asperity and the half-space becomes smaller than δ_{in}, which is calculated for each asperity. A first estimate of the asperity contact radius a_i is done by inverting the JKR relation (2). Then, after a further increment of the approach $\Delta\delta_i = z_i - w_i$, being z_i the height of the asperity i, the contact radius is increased by the quantity

$$\Delta a_i = \frac{\Delta\delta_i}{2a_i/R_i - \sqrt{\pi\Delta\gamma/(E^* a_i)}} \tag{9}$$

which is obtained by differentiating Equation (2).

The total contact area and load are then obtained by summing up the contributions of all the asperities in contact. Moreover, as a self-balanced load distribution is considered, the interfacial mean separation $\bar{u}$ is computed as $\bar{u}_0 - \delta$, where $\bar{u}_0$ and δ are the initial separation and the total approach, respectively.

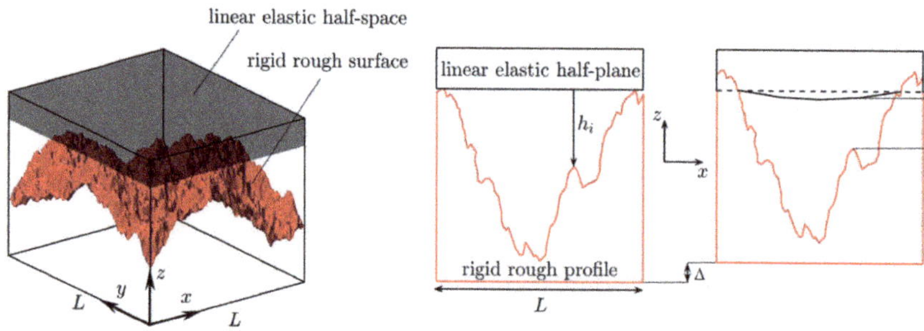

Figure 2. Elastic half-space in contact with a rigid rough surface.

3. Results

Computations have been performed on fractal self-affine isotropic surfaces. Roughness is described by its power spectral density (PSD), which has a power law relation with the magnitude $q = |\mathbf{q}|$ of the wavevector $\mathbf{q}$. In this work, we consider fractal surfaces with PSD

$$C(q) = C_0(q/q_L)^{-2(H+1)} \qquad \text{for } q_L \leq q < q_1 \qquad (10)$$

and zero otherwise. We have denoted with $q_L = 2\pi/L$ and $q_1 = 2\pi/\lambda_1$ the short and long frequencies cut-off, respectively. The quantity L represents the lateral size of the domain (in this case $L_x = L_y$). Finally, H is the so-called Hurst exponent, which is related to the fractal dimension $D_f = 3 - H$. Rough surfaces are numerically generated according to the spectral method proposed in [25,26].

In our calculations, we fixed $L = 1$ mm and $h_{rms} = 5$ µm. Furthermore, two sets of simulations have been performed. In the first one, we fixed $H = 0.8$ and $q_1 = \zeta q_L$, with magnification $\zeta = 64, 128, 256, 512$. In the second one, we fixed $\zeta = 128$ and $H = 0.45$, 0.65, 0.8, 0.95.

3.1. *Adhesive Hysteresis and Pull-Off Force: Effect of Loading Parameters*

Figure 3A shows the normalized true contact area A/A_0 as a function of the dimensionless load $F/(A_0E^*)$. Calculations have been performed for $H = 0.8$, $\zeta = 128$, and $h_{rms} = 5$ µm. Moreover, the curves are obtained by averaging the results of six surface realizations. The material properties are $E^* = 1.0$ MPa and $\Delta\gamma = 0.07$ J/m^2, which are typical values for very soft silicon elastomers. Unloading starts from different maximum applied loads $F_{max}/(E^*A_0) = 0.004, 0.0071, 0.012, 0.015$.

For adhesiveless rough contacts, the area–load relation is known to be linear [27–29]. However, recent studies confirm that adhesion may lead to strong non-linearity of the $F - A$ curve [30–32]. In our calculations, this is especially true for the unloading path in agreement with numerical [15] and experimental [33] findings. The pull-off force is the maximum negative load reached during retraction. Near the pull-off point, unloading paths almost collapse on a single curve. As a result, the pull-off force F_{po} is quite independent on the maximum true area of contact reached during the approach (Figure 3B). This is in agreement with experimental findings of Refs. [11,33].

In an approach–retraction cycle, the magnitude Θ of hysteretic losses is equal to the area between the loading and unloading $F - \Delta$ curves, where Δ is the mean penetration of the rough surface in the half-space. Figure 3C shows the evolution of the dimensionless contact force $F/(E^*A_0)$ with the normalized penetration Δ/h_{rms}. Recent experiments [5,11] suggest that, in presence of surface roughness, Θ increases linearly with the penetration Δ (or in a similar way with the true area of contact A). Such phenomenon is known in literature as depth-dependent hysteresis. Figure 3D shows the increase of the dimensionless energy loss $\Theta/(E^*A_0h_{rms})$ with A/A_0 corresponding to the maximum applied loads. The model captures the linear relation between Θ and A/A_0 and the analytical predictions given by DK's model [14], as adapted to the present case (see Appendix A), are coherent with our numerical results.

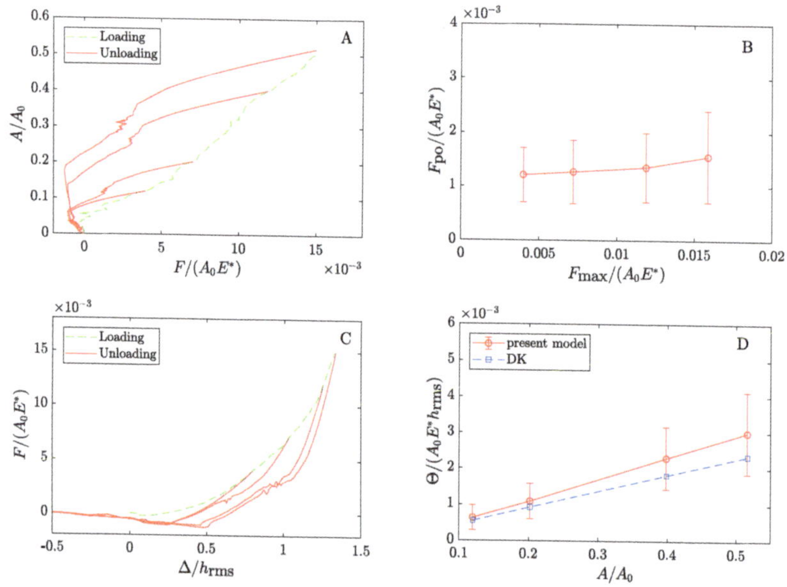

Figure 3. (**A**) The normalized area of contact A/A_0 as a function of the dimensionless load $F/(E^*A_0)$. Results are obtained on surfaces with $h_{rms} = 5$ µm, $H = 0.8$ and $\zeta = 128$. Loading (green dashed line) and unloading (red solid line) curves are shown. Results are averaged on 6 surface realizations. (**B**) The dimensionless pull-off force $F_{po}/(E^*A_0)$ as a function of the maximum applied load at the end of the loading phase. Error bars denote the standard deviation on 6 surface realizations. (**C**) The dimensionless mean penetration Δ/h_{rms} as a function of the applied load $F/(E^*A_0)$. (**D**) The dimensionless energy loss $\Theta/(E^*A_0h_{rms})$ as a function of the normalized area of contact A/A_0. Red solid and blue dashed lines refer to the present calculations and DK's predictions, respectively. Error bars denote the standard deviation on 6 surface realizations.

3.2. Adhesive Hysteresis and Pull-Off Force: Effect of the Fractal Parameters

Surface roughness can be described by its statistical parameters, i.e., RMS roughness amplitude h_{rms}, RMS gradient h'_{rms}, and RMS curvature h''_{rms}. The first one is related to low frequencies of the PSD spectrum, while RMS slope and curvature mainly depends on the cut-off frequency q_1 and therefore on the magnification ζ. Increasing ζ, the PSD spectrum is enriched by smaller and smaller roughness wavelengths. An other important parameter is the Hurst exponent H. Low (high) values of H correspond to high (low) fractal dimension D_f ($D_f = 3 - H$).

Figure 4A shows the $F/(E^*A_0) - A/A_0$ relation at increasing values of the magnification ζ ($\zeta = 64, 128, 256, 512$) for $h_{rms} = 5$ µm and $H = 0.8$. All curves are obtained for a same value of the applied load $F/(E^*A_0) = 0.071$ reached at the end of the approach, in similar way to the experiments performed in [5]. The true area of contact decreases with ζ as an increase in ζ corresponds to bigger rms gradient h'_{rms}. In such case, surface roughness is described by several length-scales and a greater load is required to create new contact patches on smaller wavelengths. Specifically, as shown in [34], the dependence of the curves on ζ (and thus on h'_{rms}) is exclusively due to the contribution of the repulsive interactions.

Figure 4B shows the dimensionless pull-off force $F_{po}/(E^*A_0)$ as a function of the magnification. A general drop in the pull-off force is observed by increasing ζ. This is in agreement with recent numerical findings in [35], where an in-house Boundary Element Method (BEM) has been developed for studying adhesive contact of rough surfaces, by including full Lennard–Jones potentials and surface integration at the asperity level. However, such results are due to the fact that we are considering surfaces with roughness amplitude of the order of microns. In fact, as found in [13], at sufficiently high values of the rms roughness amplitude (and more precisely of the product $q_0 h_{\text{rms}}$, being q_0 the roll-off frequency), for $H = 0.8$ a decrease in magnification ζ involves an increase in the effective surface energy at short length scale (large ζ). This effect results from the increase in the contact area as more a more short-wavelength roughness components are taken into account. However, we stress that such a result works as long as $q_0 h_{\text{rms}}$ is high enough. Indeed, at lower h_{rms}, such effect is not observed and the adhesion seems to be governed only by the surface roughness amplitude [36].

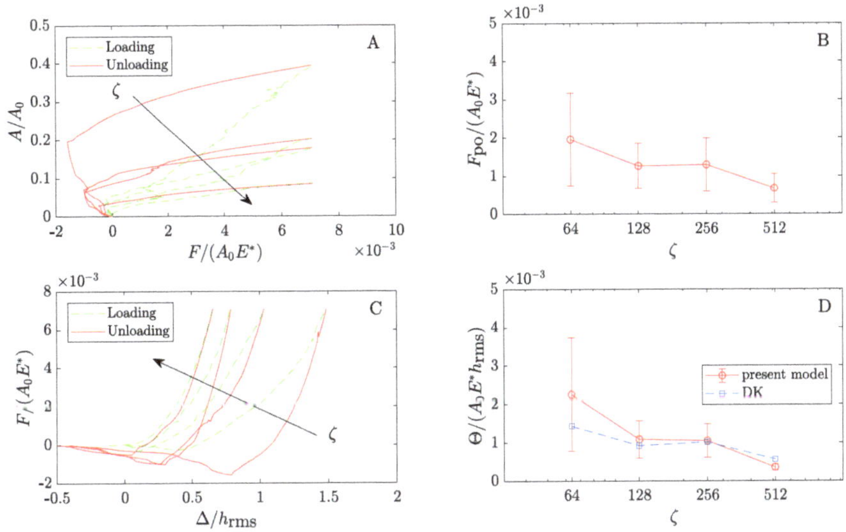

Figure 4. (**A**) The normalized area of contact A/A_0 as a function of the dimensionless load $F/(E^*A_0)$. Results are obtained on surfaces with $h_{\text{rms}} = 5$ µm, $H = 0.8$, and $\zeta = 64, 128, 256, 512$. Loading (green dashed line) and unloading (red solid line) curves are shown. Results are averaged on 6 surface realizations. (**B**) The dimensionless pull-off force $F_{\text{po}}/(E^*A_0)$ as a function of the magnification. Error bars denote the standard deviation on 6 surface realizations. (**C**) The dimensionless mean penetration Δ/h_{rms} as a function of applied load $F/(E^*A_0)$. (**D**) The dimensionless energy loss $\Theta/(E^*A_0 h_{\text{rms}})$ as a function of the magnification. Red solid and blue dashed lines refer to the present calculations and DK predictions, respectively. Error bars denote the standard deviation on 6 surface realizations.

In fact, for large ζ and small h_{rms}, a model based on a JKR-type approach becomes questionable as the dimension of the contact spots decreases [37]. In such case, a DMT-type approach, based on the assumption of long-range adhesion interactions, could be more accurate in modeling the contact problem. In this regard, a DMT-type model is developed in [34] with the aim of investigating the adhesive contact of surfaces with h_{rms} of the order of 1 nm. It is found that F_{po} is almost independent of h'_{rms}, i.e., the adhesion force required for the detachment is magnification independent.This is confirmed in [38], where a stickiness criterion [38] is derived from Persson–Scaraggi DMT theory [37]. For typical values of the Hurst exponent ($H > 0.6$), the criterion suggests that adhesion is destroyed by the long wavelengths of roughness, while ζ has negligible effects. Such result has been corroborated by very recent experimental [39] and analytical works [40,41], according to which the main parameter "killing" adhesion seems to be the roughness amplitude h_{rms}.

Figure 4C shows the load-penetration relationship for increasing magnification ζ. The corresponding hysteretic losses Θ are shown in Figure 4D. Our simulations suggest that the magnitude of energy loss follows the same trend of the pull-off force, i.e., it reduces with ζ. Once again, DK's predictions, as given by the proposed modified equation in appendix, are in agreement with our numerical calculations. The error bars in Figure 4B–D show the standard deviation on six surface realizations. The scatter is larger for surfaces with low magnification ζ as a result of the smaller number of surface asperities. On the contrary, increasing ζ, smaller and smaller asperities are added to the rough surface and their spatial distribution is expected to be more uniform in the nominal contact region. Moreover, as the linear size of the system is finite the PSD is not continuous and even assuming spectral components with random phases uniformly distributed in the range $0 < \phi < 2\pi$ the surface will be not ergodic, and a single realization of the surface will be in general highly non-Gaussian, thus entailing finite-size effects related to the finite value of the asperities heights. It follows that for surfaces without a low-wavenumber roll-off (or cut-off) region, quantities which depend on the long wavelength roughness, such as the average interfacial separation (and hence the hysteresis dissipation) at low contact pressures, will vary strongly from one realization to another.

A second set of simulations has been performed on surfaces with fixed $h_{\mathrm{rms}} = 5$ µm, $\zeta = 128$ and different Hurst exponents $H = 0.45, 0.65, 0.8, 0.95$. We have fixed again the maximum applied load $F/(E^*A_0) = 0.071$. An increase in H leads to a decrease in RMS gradient, thus explaining why the area increases with H for a fixed load (Figure 5A). Figure 5B shows that the pull-off force is destroyed at low H; the same trend has been observed in [42], where the adhesive contact between a parabolic indenter with superimposed roughness and an elastic half space has been studied in the JKR-limit. Figure 5C shows how the $F/(E^*A_0) - \Delta/h_{\mathrm{rms}}$ relation modifies with H. In particular, at $H = 0.45$ loading and unloading paths overlaps thus showing vanishing adhesive hysteretic loss Θ, which is strongly affected by the Hurst exponent as shown in Figure 5D. Notice results are more scattered for surfaces with lower fractal dimension as finite-size effects related to the absence of a low-wavenumber roll-off (or cut-off) region are exaggerated at higher values of H.

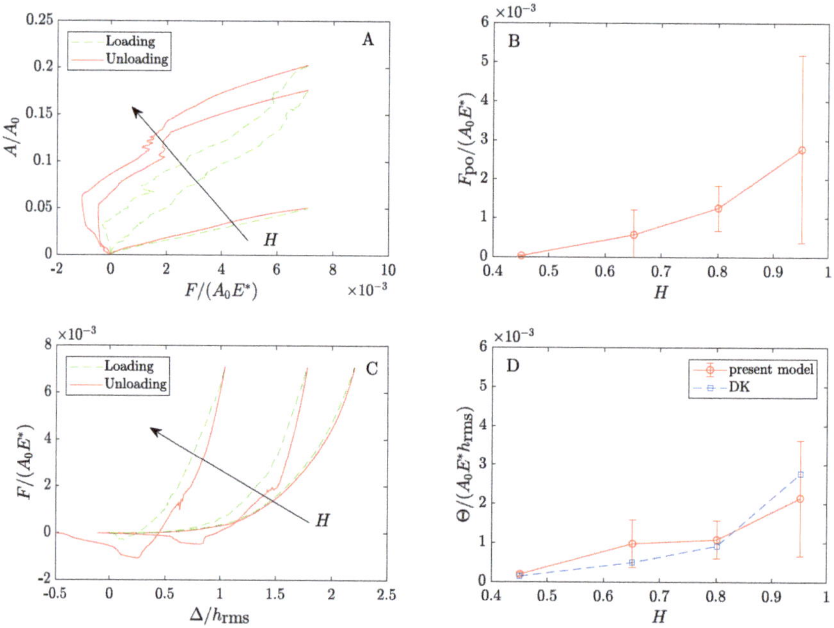

Figure 5. (**A**) The normalized area of contact A/A_0 as a function of the dimensionless load $F/(E^*A_0)$. Results are obtained on surfaces with $h_{\mathrm{rms}} = 5$ µm, $\zeta = 128$, and $H = 0.45, 0.65, 0.8$. Loading (green dashed line) and unloading (red solid line) curves are shown. Results are averaged on 6 surface realizations. (**B**) The dimensionless pull-off force $F_{\mathrm{po}}/(E^*A_0)$ as a function of the Hurst exponent. Error bars denote the standard deviation on 6 surface realizations. (**C**) The dimensionless mean penetration Δ/h_{rms} as a function of applied load $F/(E^*A_0)$. (**D**) The dimensionless energy loss $\Theta/(E^*A_0h_{\mathrm{rms}})$ as a function of the Hurst exponent. Red solid and blue dashed lines refer to the present calculations and DK predictions, respectively. Error bars denote the standard deviation on 6 surface realizations.

4. Discussion and Conclusions

In the adhesive contact between an elastic half-space and a rigid randomly rough surface, loading-unloading loops can be observed as a result of adhesion hysteresis induced by roughness. Hysteretic losses are found to be linearly increasing with the true area of contact reached at the end of the loading path. On the contrary, the pull-off force is negligibly influenced by the maximum contact area (and thus maximum applied load).

Pull-off force and hysteretic losses are strongly affected by roughness parameters. Specifically, here we have investigated the effects of the Hurst exponent H and magnification ζ.

Detachment force and hysteretic losses are observed to reduce by decreasing H and increasing ζ. Such results are related to the increase in the RMS gradient h'_{rms} occurring when H is reduced or ζ is increased. Our outcomes are in agreement with the trends shown by very recent numerical, experimental and analytical findings.

However, we stress that our results are obtained on surfaces with RMS roughness amplitude of the order of few micrometers where we can reasonably expect partial contact conditions occur in a wide range of applied loads. In fact, multiasperity models become progressively less accurate moving towards full contact conditions. Moreover, numerical models allow to consider a limited range of magnifications, while real surfaces are characterized by roughness on several length scales (with the modern technologies we can measure $\zeta \simeq 10^7$, ranging from centimeter to nanometer scales). Despite such limitations, the present findings help to clarify some aspects of the hysteretic phenomenon occurring in the adhesive contact of rough soft matter.

Author Contributions: G.V. and L.A. contributed equally to this work. All authors have read and agreed to the published version of the manuscript.

Funding: This work was supported by the project "FASTire (Foam Airless Spoked Tire): Smart Airless Tyres for Extremely-Low Rolling Resistance and Superior Passengers Comfort" funded by the Italian MIUR Progetti di Ricerca di Rilevante Interesse Nazionale (PRIN) call 2017—grant n. 2017948FEN.

Informed Consent Statement: Informed consent was obtained from all subjects involved in the study.

Data Availability Statement: The data presented in this work are available on request from the corresponding author.

Acknowledgments: L.A. and G.V. acknowledge support from the Italian Ministry of Education, University and Research (MIUR) under the program "Departments of Excellence" (L.232/2016).

Conflicts of Interest: The authors declare no conflicts of interest.

Appendix A. Deng and Kesari's Model

Deng and Kesari's model (DK) [14] is a combination of classical adhesive theories for smooth elastic spheres and Nayak's theory of rough surfaces [43]. In their work, DK used both JKR [18] and Maugis–Dugdale [44] theories for estimating the energy loss in a loading–unloading cycle of a single spherical asperity. In the limit of high Tabor parameters ($\mu \geq 3$), the two theories predict the same behavior. For this reason, as our study is focused on very soft materials, for which high Tabor numbers are expected, here we discuss DK's model in the framework of JKR theory.

DK gives an empirical estimate of the energy loss θ for a loading–unloading cycle of a single asperity of radius R; in the JKR limit the value of θ is

$$\theta = 2E^* R^3 \bar{\theta} \tag{A1}$$

being $\bar{\theta} \approx 0.5262[\pi\Delta\gamma/(E^* R)]^{5/3}$.

For a rough surface the energy loss of each asperity depends on the value of its radius of curvature. Following the work in [43], the variation of curvature in the population of all asperities contained in any unit region is

$$p_k(t) = \sqrt{\frac{3}{4\pi}}(t^2 - 2 + 2e^{-t^2/2})e^{-\frac{8(C_1^2-C_2^2)t^2}{16C_1}} \tag{A2}$$

where $t = -\sqrt{3/m_4}k_{\rm m}$, being $k_{\rm m} \in (0, \infty)$ the surface's mean curvature at the apex of an asperity, and the constants $C_1 = \alpha/(2\alpha - 3)$ and $C_2 = C_1\sqrt{12/\alpha}$ are related to the Nayak parameter $\alpha = m_0 m_4/m_2^2$. The quantities m_0, m_2, and m_4 are the spectral moments of surface roughness PSD. In particular, they are related to the rms roughness amplitude $h_{\rm rms}$, gradient $h'_{\rm rms}$ and curvature $h''_{\rm rms}$ by $h_{\rm rms} = \sqrt{m_0}$, $h'_{\rm rms} = \sqrt{2m_2}$, and $h''_{\rm rms} = \sqrt{8/3m_4}$.

Substituting (A2) in (A1) and integrating on the range of variation of t, the mean energy loss of contacting asperities can be written as

$$\langle\theta\rangle = 0.5262 \times 2E^* \int_{-\infty}^{0} dt[\pi\Delta\gamma/(E^* R(t))]^{5/3} p_k(t) R(t)^3 \tag{A3}$$

being $R(t) = (1/R_{\rm tip} - \sqrt{m_4/3}t)^{-1}$.In DK's model, $R_{\rm tip}$ is the radius of the spherical indenter that is in contact with a nominally flat surface. In our case, as contact occurs between nominally flat surfaces, $1/R_{\rm tip} \to 0$.

Finally, the total energy loss can be computed as

$$\Theta = \eta \cdot A \cdot \langle\theta\rangle \tag{A4}$$

where $\eta = m_4/(6\pi\sqrt{3}m_2)$ is the asperity density in a nominal contact region of unit area.

In the original DK's model, the contact area A is computed applying JKR theory to the macroscopic spherical indenter of radius $R_{\rm tip}$ and neglecting roughness contribution. In particular, A is calculated as $A_{\Delta\,\rm max} - A_{\Delta\rm in}$, where $A_{\Delta\,\rm max}$ and $A_{\Delta\rm in}$ are the values of the contact area at the maximum indentation and the macroscopic jump-in instability of the spherical tip, respectively.

In this work, as we are dealing with the contact between two nominally flat surfaces, $A_{\Delta\rm in}$ is interpreted as the true contact area of the first few asperities jumping into contact, while $A_{\Delta\,\rm max}$ is the true area of contact obtained at the end of the loading phase. Moreover, as in our calculations $A_{\Delta\rm in} \ll A_{\Delta\,\rm max}$, we have assumed $A_{\Delta\,\rm max} - A_{\Delta\rm in} \approx A_{\Delta\,\rm max}$.

References

1. Tiwari, A.; Dorogin, L.; Bennett, A.I.; Schulze, K.D.; Sawyer, W.G.; Tahir, M.; Heinrich, G.; Persson, B.N.J. The effect of surface roughness and viscoelasticity on rubber adhesion. *Soft Matter* **2017**, *13*, 3602–3621. [CrossRef]
2. Violano, G.; Afferrante, L. Adhesion of compliant spheres: An experimental investigation. *Procedia Struct. Integr.* **2019**, *24*, 251–258. [CrossRef]
3. Kim, J.Y.; Baltazar, A.; Rokhlin, S.I. Ultrasonic assessment of rough surface contact between solids from elastoplastic loading–unloading hysteresis cycle. *J. Mech. Phys. Solids* **2004**, *52*, 1911–1934. [CrossRef]
4. Guduru, P.R.; Bull, C. Detachment of a rigid solid from an elastic wavy surface: Experiments. *J. Mech. Phys. Solids* **2007**, *55*, 473–488. [CrossRef]
5. Dalvi, S.; Gujrati, A.; Khanal, S.R.; Pastewka, L.; Dhinojwala, A.; Jacobs, T.D. Linking energy loss in soft adhesion to surface roughness. *Proc. Natl. Acad. Sci. USA* **2019**, *116*, 25484–25490. [CrossRef]
6. Pinnaratip, R.; Bhuiyan, M.S.A.; Meyers, K.; Rajachar, R.M.; Lee, B.P. Multifunctional biomedical adhesives. *Adv. Healthc. Mater.* **2019**, *8*, 1801568. [CrossRef] [PubMed]
7. Kinloch, A.J. *Adhesion and Adhesives: Science and Technology*; Springer Science & Business Media: New York, NY, USA, 2012.
8. Fiorello, I.; Tricinci, O.; Naselli, G.A.; Mondini, A.; Filippeschi, C.; Tramacere, F.; Mazzolai, B. Climbing Plant-Inspired Micropatterned Devices for Reversible Attachment. *Adv. Funct. Mater.* **2020**, *30*, 2003380. [CrossRef]
9. Lorenz, B.; Krick, B.A.; Mulakaluri, N.; Smolyakova, M.; Dieluweit, S.; Sawyer, W.G.; Persson, B.N.J. Adhesion: Role of bulk viscoelasticity and surface roughness. *J. Phys. Condens. Matter* **2013**, *25*, 225004. [CrossRef] [PubMed]
10. Persson, B.N.J.; Tosatti, E. The effect of surface roughness on the adhesion of elastic solids. *J. Chem. Phys.* **2001**, *115*, 5597–5610. [CrossRef]

11. Kesari, H.; Doll, J.C.; Pruitt, B.L.; Cai, W.; Lew, A.J. Role of surface roughness in hysteresis during adhesive elastic contact. *Philos. Mag. Philos. Mag. Lett.* **2010**, *90*, 891–902. [CrossRef]
12. Deng, W.; Kesari, H. Molecular statics study of depth-dependent hysteresis in nano-scale adhesive elastic contacts. *Model. Simul. Mater. Sci. Eng.* **2017**, *25*, 055002. [CrossRef]
13. Persson, B.N.J. Adhesion between elastic bodies with randomly rough surfaces. *Phys. Rev. Lett.* **2002**, *89*, 245502. [CrossRef] [PubMed]
14. Deng, W.; Kesari, H. Depth-dependent hysteresis in adhesive elastic contacts at large surface roughness. *Sci. Rep.* **2019**, *9*, 1–12. [CrossRef]
15. Carbone, G.; Pierro, E.; Recchia, G. Loading-unloading hysteresis loop of randomly rough adhesive contacts. *Phys. Rev. E* **2015**, *92*, 062404. [CrossRef]
16. Wei, Z.; He, M.F.; Zhao, Y.P. The effects of roughness on adhesion hysteresis. *J. Adhes. Sci. Technol.* **2010**, *24*, 1045–1054. [CrossRef]
17. Greenwood, J.A. Reflections on and extensions of the Fuller and Tabor theory of rough surface adhesion. *Tribol. Lett.* **2017**, *65*, 159. [CrossRef]
18. Johnson, K.L.; Kendall, K.; Roberts, A. Surface energy and the contact of elastic solids. *Proc. R. Soc. Lond. A Math. Phys. Sci.* **1971**, *324*, 301–313.
19. Ciavarella, M.; Delfine, V.; Demelio, G. A "re-vitalized" Greenwood and Williamson model of elastic contact between fractal surfaces. *J. Mech. Phys. Solids* **2006**, *54*, 2569–2591. [CrossRef]
20. Afferrante, L.; Carbone, G.; Demelio, G. Interacting and coalescing Hertzian asperities: A new multiasperity contact model. *Wear* **2012**, *278*, 28–33. [CrossRef]
21. Wu, J.J. The jump-to-contact distance in atomic force microscopy measurement. *J. Adhes.* **2010**, *86*, 1071–1085. [CrossRef]
22. Tabor, D. Surface Forces and Surface Interactions. In *Plenary and Invited Lectures*; Academic Press: Cambridge, MA, USA, 1977; pp. 3–14.
23. Ciavarella, M.; Greenwood, J.A.; Barber, J.R. Effect of Tabor parameter on hysteresis losses during adhesive contact. *J. Mech. Phys. Solids* **2017**, *98*, 236–244. [CrossRef]
24. Violano, G.; Afferrante, L. Modeling the adhesive contact of rough soft media with an advanced asperity model. *Tribol. Lett.* **2019**, *67*, 119. [CrossRef]
25. Putignano, C.; Afferrante, L.; Carbone, G.; Demelio, G. A new efficient numerical method for contact mechanics of rough surfaces. *Int. J. Solids Struct.* **2012**, *49*, 338–343. [CrossRef]
26. Putignano, C.; Menga, N.; Afferrante, L.; Carbone, G. Viscoelasticity induces anisotropy in contacts of rough solids. *J. Mech. Phys. Solids* **2019**, *129*, 147–159. [CrossRef]
27. Afferrante, L.; Bottiglione, F.; Putignano, C.; Persson, B.N.J.; Carbone, G.J.T.L. Elastic contact mechanics of randomly rough surfaces: An assessment of advanced asperity models and Persson's theory. *Tribol. Lett.* **2018**, *66*, 75. [CrossRef]
28. Putignano, C.; Afferrante, L.; Carbone, G.; Demelio, G. The influence of the statistical properties of self-affine surfaces in elastic contacts: A numerical investigation. *J. Mech. Phys. Solids* **2012**, *60*, 973–982. [CrossRef]
29. Muser, M.H.; Dapp, W.B.; Bugnicourt, R.; Sainsot, P.; Lesaffre, N.; Lubrecht, T.A.; Persson, B.N.J.; Harris, K.; Bennett, A.; Schulze, K.; et al. Greenwood Meeting the contact-mechanics challenge. *Tribol. Lett.* **2017**, *65*, 118. [CrossRef]
30. Salehani, M.K.; van Dokkum, J.S.; Irani, N.; Nicola, L. On the load-area relation in rough adhesive contacts. *Tribol. Int.* **2020**, *144*, 106099. [CrossRef]
31. Violano, G.; Afferrante, L. Contact of rough surfaces: Modeling adhesion in advanced multiasperity models. *Proc. Inst. Mech. Eng. Part J. Eng. Tribol.* **2019**, *233*, 1585–1593. [CrossRef]
32. Violano, G.; Afferrante, L. On DMT methods to calculate adhesion in rough contacts. *Tribol. Int.* **2019**, *130*, 36–42. [CrossRef]
33. Krick, B.A.; Vail, J.R.; Persson, B.N.; Sawyer, W.G. Optical in situ micro tribometer for analysis of real contact area for contact mechanics, adhesion, and sliding experiments. *Tribol. Lett.* **2012**, 45, 185–194. [CrossRef]
34. Violano, G.; Demelio, G.; Afferrante, L. A note on the effect of surface topography on adhesion of hard elastic rough bodies with low surface energy. *J. Mech. Behav. Mater.* **2019**, *28*, 8–12. [CrossRef]
35. Ghanbarzadeh, A.; Faraji, M.; Neville, A. Deterministic normal contact of rough surfaces with adhesion using a surface integral method. *Proc. R. Soc. A* **2020**, *476*, 20200281. [CrossRef]
36. Persson, B.N.J.; Sivebæk, I.M.; Samoilov, V.N., Zhao, K.; Volokitin, A.I.; Zhang, Z. On the origin of Amonton's friction law. *J. Phys. Condens. Matter* **2008**, *20*, 395006. [CrossRef]
37. Persson, B.N.; Scaraggi, M. Theory of adhesion: Role of surface roughness. *J. Chem. Phys.* **2014**, *141*, 124701. [CrossRef] [PubMed]
38. Violano, G.; Afferrante, L.; Papangelo, A.; Ciavarella, M. On stickiness of multiscale randomly rough surfaces. *J. Adhes.* **2019**, 1–19. doi:10.1080/00218464.2019.1685384 [CrossRef]
39. Tiwari, A.; Wang, J.; Persson, B.N.J. Adhesion paradox: Why adhesion is usually not observed for macroscopic solids. *Phys. Rev. E* **2020**,*102*, 042803. [CrossRef] [PubMed]
40. Joe, J.; Thouless, M.D.; Barber, J.R. Effect of roughness on the adhesive tractions between contacting bodies. *J. Mech. Phys. Solids* **2018**, *118*, 365–373. [CrossRef]
41. Joe, J.; Scaraggi, M.; Barber, J.R. Effect of fine-scale roughness on the tractions between contacting bodies. *Tribol. Int.* **2017**, *111*, 52–56. [CrossRef]
42. Li, Q.; Pohrt, R.; Popov, V.L. Adhesive Strength of Contacts of Rough Spheres. *Front. Mech. Eng.* **2019**, *5*, 7. [CrossRef]

43. Nayak, P.R. Random Process Model of Rough Surfaces. *J. Lubr. Tech.* **1971**, *93*, 398–407. [CrossRef]
44. Maugis, D. Adhesion of spheres: The JKR-DMT transition using a Dugdale model. *J. Colloid Interface Sci.* **1992**, *150*, 243–269. [CrossRef]

lubricants

MDPI

Article

A Numerical Study on Roughness-Induced Adhesion Enhancement in a Sphere with an Axisymmetric Sinusoidal Waviness Using Lennard–Jones Interaction Law

Antonio Papangelo [1,2,*] and Michele Ciavarella [1,2]

[1] Politecnico di BARI, Department of Mechanics, Mathematics and Management, Via Orabona 4, 70126 Bari, Italy; mciava@poliba.it

[2] Department of Mechanical Engineering, Hamburg University of Technology, Am Schwarzenberg-Campus 1, 21073 Hamburg, Germany

* Correspondence: antonio.papangelo@poliba.it

Received: 3 August 2020; Accepted: 3 September 2020; Published: 4 September 2020

Abstract: Usually, roughness destroys adhesion and this is one of the reasons why the "adhesion paradox", i.e., a "sticky Universe", is not real. However, at least with some special type of roughness, there is even the case of adhesion enhancement, as it was shown clearly by Guduru, who considered the contact between a sphere and a wavy axisymmetric single scale roughness, in the limit of short range adhesion (JKR limit). Here, the Guduru's problem is numerically solved by using the Boundary Element Method (BEM) with Lennard–Jones interaction law, which allowed us to explore the contact solution from the rigid to the JKR limit. It is shown that adhesion enhancement stops either for low Tabor parameter, or by large waviness amplitudes, due to the appearance of internal cracks within the contact patch. We do not seem to find a clear threshold for "stickiness" (complete elimination of adhesion), contrary to other recent theories on random roughness. The enhancement effect is well captured by an equation in terms of the Johnson parameter derived by Ciavarella–Kesari–Lew, and is much larger than the Persson–Tosatti enhancement in terms of increase of real contact area due to roughness. The Persson–Tosatti energetic argument for adhesion reduction seems to give a lower bound to the effective work of adhesion.

Keywords: adhesion; roughness; adhesion enhancement; JKR model; Lennard–Jones

1. Introduction

Adhesion is a challenging topic in tribology [1–3] with relevance in several engineering applications that range from biomimetics [4], soft matters [5], soft robots [6], grippers [7], friction [8–12]. Although roughness is usually responsible for adhesion reduction [13–15], Briggs and Briscoe [16] showed already in 1977 that relatively small random roughness amplitude could enhance adhesion in pull-off experiments as well as relative rolling resistance by a factor up to 2.5. Later, Guduru [17] showed that in the contact between a rigid sphere and a soft halfspace with an axisymmetric single wavelength waviness, adhesion could be enhanced by a factor up to 20 with respect to the Johnson–Kendall–Roberts smooth case ([18], JKR in the following). The enhancement was first modeled theoretically by Guduru [17] and then proved experimentally by Guduru and Bull [19]. The basic assumptions of the Guduru [17] model are that (i) the contact area is simply connected (there are no circular grooves within the contact patch) and that (ii) the halfspace is constituted by a soft material (elastomer or rubber) hence adhesion can be simply modeled by JKR theory [18]. Loading and unloading a rigid sphere from the wavy surface leads to several jump instabilities and related dissipation, which is responsible for the measured enhancement. Kesari et al. [20] showed that if the roughness wavelength is substantially shorter than the sphere radius, then an envelope solution can be obtained, which describes well the loading-unloading hysteretical behavior well known to experimentalists (see also Kesari and Lew [21]).

Waters and coauthors in [22] developed a Maugis–Dugdale cohesive model, still based on the assumption of simply connected contact area, to account for the transition between the rigid and JKR limit. They showed that toughening and strengthening of the interface was mostly restricted to the JKR regime, while, in the rigid limit, they found the Bradley [23] solution for the smooth rigid sphere. Ciavarella [24] further discussed the assumptions of the Guduru model and the conclusions of Waters and coauthors [22]. In particular he noticed that for hard solids (i.e., in the rigid limit) the axisymmetric roughness should reduce the macroscopic adhesion by orders of magnitude with respect to the smooth sphere limit. Ciavarella [24] supported his argument by considering the Rumpf–Rabinowich model ([25–27]), which geometry is analogous but not equal to that of Guduru and is used for adhesion of hard particles (the model neglects the elastic deformation). The Rumpf–Rabinowich model predicts that increasing the substrate roughness the macroscopic adhesion force first decreases and then increases again. Ciavarella [24] suggested that the Guduru and the Rumpf–Rabinowich models may be respectively close to an upper and a lower bound for macroscopic adhesion of rough bodies (see also Ciavarella [28]).

In this paper, we reconsider the geometry of Guduru [17] and obtain a closed form solution for the rigid limit, which clearly shows that increasing the waviness amplitude A reduces the macroscopic adhesion force by orders of magnitude. By using the axisymmetric Boundary Element Method (BEM) the contact problem is solved with Lennard–Jones interaction law, for varying waviness amplitude A and wavelength λ and for different Tabor parameters of the sphere μ, without the restrictive assumption of a compact contact area. Numerical results are well in agreement with the theory both in the rigid and in the JKR limit. The transition from one regime to the other is numerically studied using the BEM code. In the JKR regime adhesion enhancement is well captured by the Johnson parameter as derived by Ciavarella–Kesari–Lew [21,24], and is much larger than the Persson–Tosatti enhancement [13] in terms of increase of real contact area due to roughness. It is shown that at large Tabor parameters μ (> 3), increasing A first leads to adhesion enhancement as predicted by Guduru theory [17], but then strongly reduces the macroscopic adhesive force due to the appearance of internal cracks. We found that for $A/\lambda \gtrsim 10^{-1}$ the JKR solution greatly overestimates the pull-off force and the hysteretical dissipation.

2. Guduru Contact Problem

JKR Theory

Guduru [17] considered the contact between a rigid sphere of radius R that indents and elastic halfspace (Young modulus E, Poisson ratio ν) with an axisymmetric waviness of wavelength λ and amplitude A (see Figure 1).

In the system of reference shown in Figure 1, the axisymmetric waviness has the form

$$y(r) = -A\left(1 - \cos\left(\frac{2\pi r}{\lambda}\right)\right) \tag{1}$$

where r is the radial coordinate. Using the Hertzian approximation for the spherical profile [29] one gets the gap function

$$f(r) = \frac{r^2}{2R} + A\left(1 - \cos\left(\frac{2\pi r}{\lambda}\right)\right) \tag{2}$$

Guduru [17] solved the adhesive contact problem under the assumptions of compact contact area (i.e., there are no axisymmetric grooves within the contact patch) and in the limit of short range adhesion [18], which requires soft bodies into contact with large surface energy. Ensuring the contact area is compact requires that the gap function is strictly monotonically increasing

$$\frac{df(r)}{dr} > 0, \qquad r > 0 \tag{3}$$

which for the gap function (2) implies

$$\frac{\lambda^2}{AR} > 8.5761 \tag{4}$$

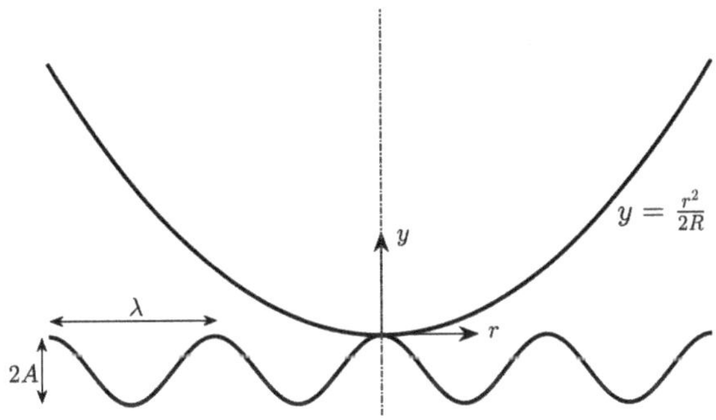

Figure 1. The geometry of the axisymmetric contact problem. A rigid sphere of radius R indents an elastic halfspace with an axisymmetric waviness of wavelength λ and amplitude A. The sphere is approximated by a Hertzian profile.

Nevertheless, condition (4) is too restrictive. Indeed, Guduru [17] analysis holds at detachment if one requires that the normal load is increased from 0 to a value such that the contact radius a gets larger than a critical radius $r_c = 2\pi AR/\lambda$ for which the gap function is strictly monotone and any partial contact within the contact patch has coalesced. To this end it is evident from Johnson [30] analysis (strictly speaking that was a 2D problem) that a simply connected contact area would be achieved also when condition (4) is violated provided that the so-called "Johnson parameter" $\alpha_{KLJ} = \frac{2\lambda w_c}{\pi^2 A^2 E^*}$ is sufficiently high to ensure spontaneous snap into full contact. By using three different solution approaches Guduru [17] obtained that the JKR adhesive solution for the geometry in Figure 1 can be written in dimensionless form as

$$\begin{cases} \overline{W}_1 = 4\beta\left[\frac{2\overline{a}^3}{3} + \alpha\left(\frac{4\pi^2\overline{a}^3}{3} + \frac{\pi\overline{a}}{2}H_1(2\pi\overline{a}) - \pi^2\overline{a}^2H_2(2\pi\overline{a})\right)\right] \\ \overline{W} = \overline{W}_1 - 4\sqrt{\beta\overline{a}^3} \\ \overline{\Delta} = \overline{a}^2 + \alpha\pi^2\overline{a}H_0(2\pi\overline{a}) - \sqrt{\frac{\overline{a}}{\beta}} \end{cases} \tag{5}$$

where the following dimensionless parameters have been defined

$$\alpha = \frac{AR}{\lambda^2}, \qquad \beta = \frac{\lambda^3 E^*}{2\pi w_c R^2}, \qquad \overline{W} = \frac{W}{\pi w_c R}, \qquad \overline{\Delta} = \frac{\Delta R}{\lambda^2}, \qquad \overline{a} = \frac{a}{\lambda}, \tag{6}$$

and $H_n(\cdot)$ is the Struve function of order n, $E^* = E/\left(1-\nu^2\right)$ is the composite elastic modulus, w_c is the surface energy per unit area, W is the external load, W_1 is the normal load in the adhesiveless problem and Δ is the remote approach (>0 when the punch approaches the halfspace, see also [22]). Inspection of Equation (5) reveals that the Guduru problem, in the JKR regime, depends on two dimensionless parameters $\{\alpha, \beta\}$: α represents the degree of waviness of the surface, with large (small)

α implying high (low) amplitude waviness, while β can be interpreted as the ratio between the elastic and the surface energy, with large (small) β implying a stiff (compliant) material [22]. For $\alpha = 0$ the classical sphere-flat JKR solution is retrieved

$$\overline{W} = 4\beta\frac{2\overline{a}^3}{3} - 4\sqrt{\beta\overline{a}^3} \tag{7}$$

$$\overline{\Delta} = \overline{a}^2 - \sqrt{\frac{\overline{a}}{\beta}} \tag{8}$$

3. Numerical Solution

3.1. Axisymmetric BEM Formulation

In the previous section the JKR solution of the Guduru contact problem was briefly summarized. Several variants of the Guduru contact problem have been studied by different authors [20–22,24,28], nevertheless all of them assume full contact within the contact patch (a simply connected contact area). To overcome this limitation an axisymmetric Boundary Element Method (BEM) was developed assuming that the rigid sphere and the wavy halfspace interacts with a Lennard–Jones interaction law (LJ in the following, see Figure 2a)

$$\sigma(h) = -\frac{8w_c}{3\varepsilon}\left[\left(\frac{\varepsilon}{h}\right)^3 - \left(\frac{\varepsilon}{h}\right)^9\right] \tag{9}$$

where σ is the traction ($\sigma > 0$, when compressive), h is the gap and ε the equilibrium distance (the maximum tensile stress $\sigma_0 = -\frac{16w_c}{9\sqrt{3}\varepsilon}$ takes place at separation $h = 3^{1/6}\varepsilon$). BEM contact codes that use the LJ interaction law have been derived previously by several authors to solve contact problems similar to the one tackled here. For example Wu [31] solved the adhesive contact problem between a sphere and a longitudinal wavy surface, while Medina and Dini [32] studied the problem of an adhesive sphere squeezed against a rough substrate. Notice that other BEM solution strategies exist that are based on energy minimization [33,34].

The contact problem considered is equivalent to the case of a "rough" axisymmetric rigid Hertzian indenter squeezed against an elastic halfspace (Figure 2b). The Guduru gap function is written as

$$h = -\Delta + \varepsilon + \frac{r^2}{2R} + A\left(1 - \cos\left(\frac{2\pi r}{\lambda}\right)\right) + u_z(r) \tag{10}$$

where $\Delta > 0$ when the Hertzian profile approaches the halfspace and $u_z(r)$ is the deflection of the elastic halfspace (see Figure 2b).

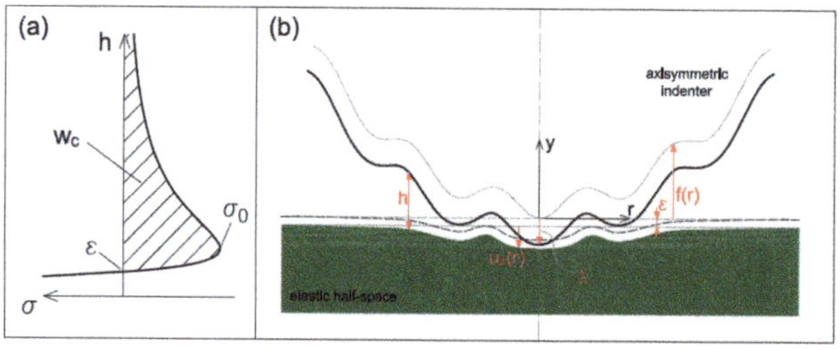

Figure 2. (**a**) Lennard–Jones interaction law. (**b**) Equivalent representation of the considered contact problem in the undeformed configuration (grey lines) and after applying a certain remote displacement Δ (black lines).

For axisymmetric frictionless contact problems [35,36]

$$u_z(r) = \frac{1}{E^*}\int \sigma(s)\, G(r,s)\, s ds \tag{11}$$

where $\sigma(s)$ is the pressure distribution, $G(r,s)$ the Kernel function

$$G(r,s) = \begin{cases} \frac{4}{\pi r} K\left(\frac{s}{r}\right), & s < r \\ \frac{4}{\pi s} K\left(\frac{r}{s}\right), & s > r \end{cases} \tag{12}$$

and $K(k)$ the complete elliptic integral of the first kind of modulus k.

After surveying the Literature on the topic [31–37], we developed an axisymmetric BEM inspired by the works of Greenwood [35] and Feng [36]. Assume the radial domain is discretized with N elements, so that we have $M = N + 1$ discretization points. To solve Equation (10) on a discrete domain, one needs to determine the elastic deflection $u_z(r)$. A problem arises in evaluating the integral (11) as the kernel function $G(r,s)$ is singular in $s = r$. The common approach is to discretize Equation (11) assuming that the pressure $\sigma(s)$ has a simple form over a discrete element. To this end, the simplest approach is to assume that the pressure is constant over each element. Nevertheless, Greenwood [35] reported that this method may lead to suspicious results, particularly in the regions with strong pressure gradients and suggested using the method of the overlapping triangles [29], for which the pressure $\sigma(s)$ has a triangular form. Hence the deflection at point r_i due to a triangular pressure distribution being p_j at $r = r_j$ and falling linearly to 0 at $r = r_{j-1}$ and $r = r_{j+1}$ is

$$u_z(r_i) = u_{z,i} = \frac{1}{E^*} G_{ij} p_j \tag{13}$$

where we have solved numerically the integral in Equation (11) to obtain G_{ij} once for all. Notice that the kernel function singularity at $r_i = r_{j-1}$ and $r_i = r_{j+1}$ is canceled by the pressure being 0 in r_{j-1} and r_{j+1}, instead, for the singular point $r_i = r_j$, we considered a pressure equal to $-p_j$ at r_{j-1} and r_{j+1} rising linearly at 0 at r_j superimposed to a constant pressure ring, equal to p_j, in between the radii r_{j-1} and r_{j+1} for which the displacement field is known analytically (see Appendix A). By defining the following quantities

$$H = h/\varepsilon - 1; \qquad u = \frac{r}{\Gamma}; \qquad L = \frac{\lambda}{\Gamma}; \qquad P = \frac{p\mu\varepsilon}{w_c}; \qquad \Gamma = \left(\frac{R^2 w_c}{E^*}\right)^{1/3}; \tag{14}$$

Equation (10) is written for the normalized gap H_i (H vanishes for $h = \varepsilon$) at each node i as:

$$H_i = -\Delta^\dagger + \frac{1}{2}\mu u_i^2 + A^\dagger\left(1 - \cos\left(\frac{2\pi u}{L}\right)\right) + \mu \sum_{j=1}^{N} G'_{ij} P_j \tag{15}$$

where $\mu = \left(\frac{R w_c^2}{E^{*2}\varepsilon^3}\right)^{1/3}$ is the Tabor parameter, $\Delta^\dagger = \Delta/\varepsilon$, $A^\dagger = A/\varepsilon$, $G'_{ij} = G_{ij}/\Gamma$ and

$$P_j = -\frac{8}{3}\mu\left[\frac{1}{(H_j+1)^3} - \frac{1}{(H_j+1)^9}\right] \tag{16}$$

All the results that will be presented below were obtained using $N = 500$ discretization elements with a constant element size, which proved to be sufficient for obtaining converged solutions up to a Tabor parameter $\mu = 5$. Lower values of μ did not require such a mesh refinement, nevertheless, to avoid confusion, the same discretization was used along all the paper. Unless differently stated, in all the simulations the overall domain length was set equal to the sphere radius $r_{\max} = R = 100\varepsilon$. The numerical code was implemented in MATLAB. Similarly to Feng [36] an efficient pseudo-archlength continuation scheme was implemented [38], which is needed to follow the system solution branches. Furthermore, to make the numerical solution of the nonlinear system of Equation (15) faster to solve, the system Jacobian was provided analytically to the numerical solver ("*fsolve*" in MATLAB) that implements a Newton–Raphson scheme.

3.2. Validation of the Numerical Results

First to assess the correctness of the numerical implementation, the BEM results are validated against those provided by Feng [36]. In Figure 3 the dimensionless normal load

$$\overline{W} = \frac{W}{\pi w_c R} = \frac{2\Gamma^2}{\mu R \varepsilon} \int_0^{+\infty} P(u) u du \tag{17}$$

is plotted as a function of the dimensionless approach $-\Delta^{\dagger}$ for $\mu = [1, 2, 3]$. Markers refer to the data reported by Feng [36] (its Figure 3) while the solid black lines were obtained numerically using the BEM presented here. Red dots stand for the Bradley [23] rigid solution, which is compared to the numerical solution (solid black line) obtained with $\mu = 0.01$. All the curves are in perfect agreement.

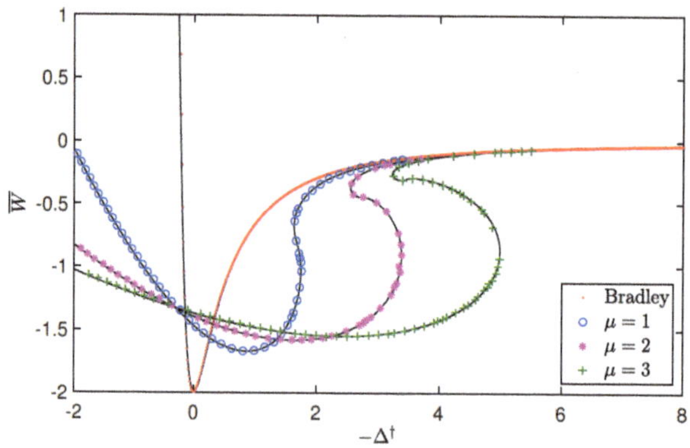

Figure 3. Dimensionless normal load $\overline{W}$ as a function of the approach $-\Delta^{\dagger}$ for $\mu = [1, 2, 3]$ as reported by Feng [36] (markers) and as obtained here numerically (solid black line). Red dots show the Bradley [23] rigid solution, which is compared to the numerical solution (solid black line) obtained with $\mu = 0.01$.

Figure 4 shows the pull-off force $|\overline{W}|_{pull-off}$ (panel (a)) and the approach $-\Delta^{\dagger}|_{pull-off}$ (panel (b)) at pull-off as a function of μ (the pull-off force $\overline{W}_{pull-off}$ is defined as the minimum of the $W(\Delta^{\dagger})$ curve). Blue stars have been obtained from Feng [36], while red squares have been obtained using the BEM developed here. The results we obtained for both load and approach at pull-off compare very favorably with those in [36].

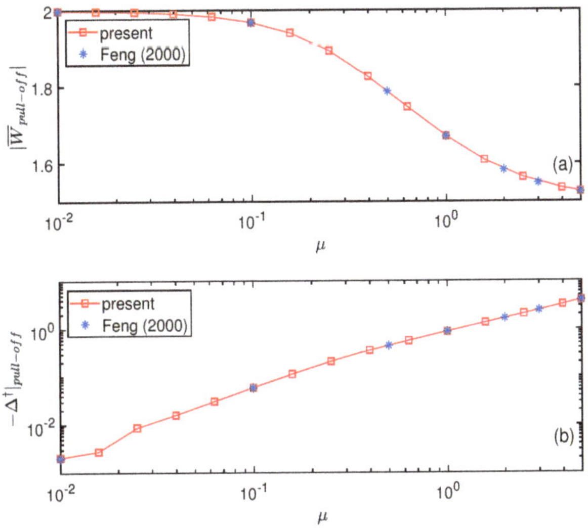

Figure 4. (**a**) Pull-off force $|\overline{W}|_{pull-off}$ and (**b**) approach $-\Delta^{\dagger}|_{pull-off}$ at pull-off as a function of the Tabor parameter $\mu = [10^{-2}, 5]$. Red squares were obtained with the presented Boundary Element Method (BEM) scheme, while blue asterisks were obtained from Feng [36].

4. Rigid Limit

The majority of the authors, who have tackled the Guduru contact problem, have focused their attention on the JKR limit, where it was clear since the early papers by Guduru [17] and Guduru and Bull [19] that substantial enhancement could be obtained, with few exceptions, as the works of Waters [22] and that of Ciavarella [24]. Waters et al. [22] developed a Maugis–Dugdale cohesive model for the Guduru problem that showed adhesion enhancement is mostly limited to the JKR regime. The cohesive model clearly depended on an additional parameter with respect to the JKR model, the Tabor parameter μ. Strictly speaking, Waters et al. [22] used the parameter introduced by Maugis [39], which anyway differs only by a small multiplicative factor from μ. Waters et al. [22] analysis showed that for small μ the pull-off detachment force converged to the Bradley rigid solution for the sphere, i.e., $|\overline{W}_B| = 2$. Nevertheless, this holds only for a smooth sphere in contact with a flat halfspace. For example, let us assume $\lambda \approx R \approx A$, then the pull-off force in the rigid regime could be estimated by considering the contact between the first crest of the halfspace waviness and the sphere. The radius of curvature of the crest is

$$R_2 = \frac{\lambda^2}{4\pi^2 A} \approx \frac{R}{4\pi^2} \tag{18}$$

and the composite radius

$$R^* = \left(\frac{1}{R} + \frac{4\pi^2}{R}\right)^{-1} \approx \frac{R}{4\pi^2} \tag{19}$$

hence the pull-off force of the sphere would be reduced by about factor $1/4\pi^2 \sim 0.025$. Indeed, using the Rumpf–Rabinowicz model [25–27] Ciavarella [24] recognized this. Although analogous to the Guduru problem, the Rumpf–Rabinowicz model refers to a different geometry. It considers the contact between a large sphere of radius R and a rigid small hemisphere of radius r_2 placed on a rigid plane (see Figure 5).

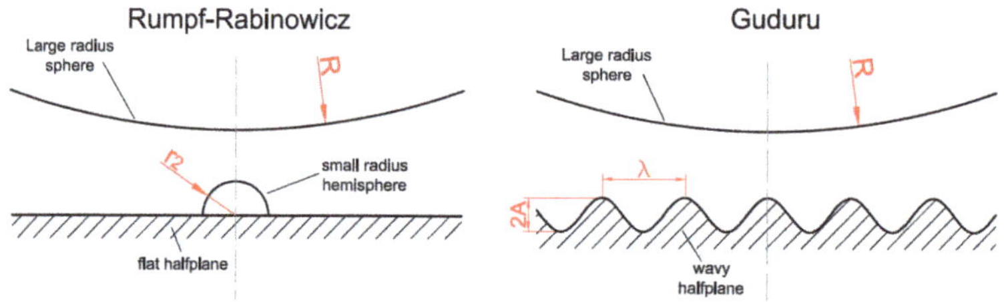

Figure 5. Left panel: simplified sketch of the Rumpf–Rabinowicz model. Right panel: simplified sketch of the Guduru model.

Two competitive mechanisms for adhesion take place: while the radius of the hemisphere increases macroscopic adhesion increases due to the interaction with the hemisphere but decreases as the rigid plane gets further away from the countersurface, which, using $r_2 = \frac{\lambda^2}{4\pi^2 A}$, can be written as [24]

$$\left|\overline{W}\right|_{pull-off} = \frac{1}{1+4\pi^2\alpha} + \frac{1}{\left(1+\frac{R^\dagger}{4\pi^2\alpha}\right)^2} \tag{20}$$

where $R^\dagger = R/\varepsilon$. Similar mechanisms are expected to be at play in the Guduru problem.

Assume that the rigid sphere and the wavy halfspace interact with a Lennard–Jones interaction law (9). From Equation (15), neglecting the elastic deformations, the dimensionless interfacial gap is

$$H(\theta) = -\Delta^\dagger + A^\dagger \left[\frac{\theta^2}{8\pi^2\alpha} + (1-\cos(\theta))\right] \tag{21}$$

where $\theta = \frac{2\pi r}{\lambda}$ has been introduced. Using Equations (16) and (17) the total load is

$$\overline{W}_{rigid} = \frac{2\pi}{\pi w_c R} \int_0^{+\infty} \sigma(H) r dr = -\frac{4A^\dagger}{3\pi^2\alpha} \int_0^{+\infty} \left[\left(\frac{1}{H(\theta)+1}\right)^3 - \left(\frac{1}{H(\theta)+1}\right)^9\right] \theta d\theta \tag{22}$$

which clarifies that at a given approach $\Delta^\dagger$ the rigid solution depends only on two parameters $\{\alpha, A^\dagger\}$.

Notice that for a smooth sphere-flat contact one can use in the first integral of Equation (22) $dH = \frac{r}{R}dr$ and obtain that the rigid solution depends only on the adhesion energy and not on the shape of the potential. This cannot be done for the Guduru gap function, which implies that the Guduru rigid solution will be slightly affected by the shape of the interaction law used.

In Figure 6 the loading curves are shown for $\log_{10}\alpha = [-4, -2, -1, 0, 1]$ and $A^\dagger = 1$. The solid black lines are the theoretical predictions based on Equation (22), while the red markers have been obtained numerically using the BEM with $\mu = 10^{-4}$. Numerical and theoretical predictions are in perfect agreement, with the curve $\log_{10}\alpha = -4$ corresponding to the Bradley [23] rigid solution for the smooth sphere (Equation (22))

$$\overline{W}_B\left(\Delta^\dagger\right) = -2\left(\frac{4\left(\Delta^\dagger-1\right)^6 - 1}{3\left(\Delta^\dagger-1\right)^8}\right) \tag{23}$$

The curves dimensionless normal load $\overline{W}$ versus dimensionless approach $-\Delta^\dagger$ plotted in Figure 6 show that the pull-off force is not monotonically decreasing with α and that the critical approach at detachment is close to $\Delta^\dagger \simeq 0$ only for the smallest values of α. Figure 7 shows the pull-off force as a function of α for $\log_{10} A^\dagger = [-3, -0.5, 0, 1, 2, 4]$ (solid lines obtained as the minimum of (22), while red markers obtained numerically $(\mu = 10^{-4})$). All the curves start at small α from the Bradley rigid solution $\left|\overline{W}_{B,pull-off}\right| = 2$ (smooth sphere), while increasing α the pull-off force decays and, after a transition zone around $\alpha \approx 1$, reaches a limit value at large α. It is shown that at large α

and $A^\dagger$ the pull-off force can decrease by more than three orders of magnitude with respect to the smooth case. To allow a comparison the predictions of the Rumpf–Rabinowicz model are reported for $\log_{10}(R^\dagger) = [0,1,2,3,4]$ (blue dashed lines in Figure 7). First notice that, while the rigid limit of the Guduru problem depends on $\{\alpha, A^\dagger\}$, the Rumpf-Rabinowicz model depends on $\{\alpha, R^\dagger\}$. Both the models show a similar decay with α, but they give two different limits for large α. In the Rumpf-Rabinowicz model, a large α implies a very small hemisphere ($R/r_2 = 4\pi^2\alpha$), hence the case of a large sphere interacting with a smooth plane is retrieved. In the rigid Guduru model, $A^\dagger$ and $\lambda^\dagger = \lambda/\varepsilon$ are not coupled, hence increasing α leads to a vanishing wavelength of the sinusoid but does not affect $A^\dagger$, which gives the observed adhesion reduction.

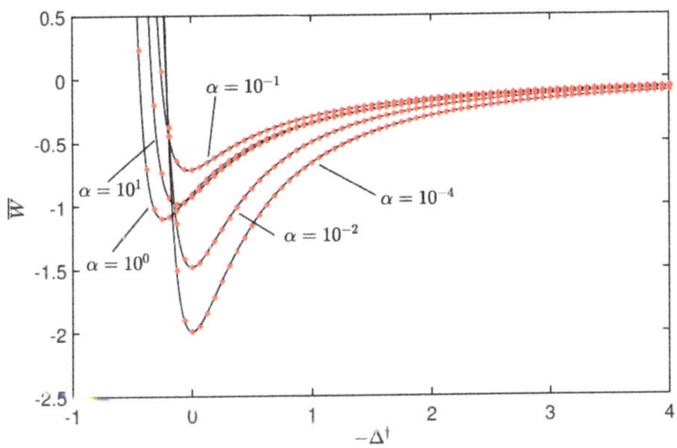

Figure 6. Loading curves for the rigid model. Solid lines have been obtained from Equation (22), while red markers are numerical solutions for $\mu = 10^{-4}$. The curves are obtained for $\log_{10}\alpha = [-4,-2,-1,0,1]$ and $A^\dagger = 1$.

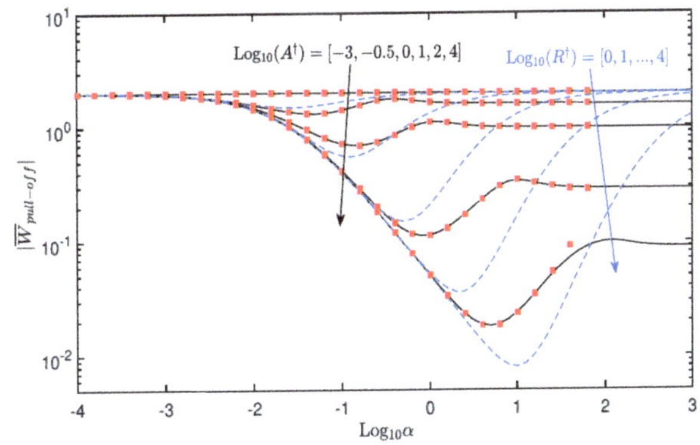

Figure 7. Pull-off force as a function of the parameter α (log scale) obtained using the rigid solution of the Guduru model (Equation (22), solid black line) and the Rumpf–Rabinowicz model (Equation (20), blue dashed line), while red squares are numerical solutions for $\mu = 10^{-4}$. For the Guduru rigid model the curves are obtained for $\log_{10} A^\dagger = [-3,-0.5,0,1,2,4]$, while for the Rumpf-Rabinowicz model $\log_{10}(R^\dagger) = [0,1,2,3,4]$.

5. Numerical Results

5.1. Effect of the Tabor Parameter

In the previous subsections we have discussed two limits of the Guduru contact problem: the JKR and the rigid limit. Here the transition from one limit to the other is investigated numerically by using the BEM introduced in Section 3.1. Figure 8 shows the pull-off force as a function of the Tabor parameter μ for $\lambda^{\dagger} = 20$, and $A^{\dagger} = [0.1, 1, 10]$. Small waviness amplitude $A^{\dagger} = 0.1$ (red circles) slightly perturbs the solution of the smooth sphere. Indeed at low Tabor parameter the pull-off force is equal to $|\overline{W}|_{pull-off} \approx 1.75$, while at higher μ it gets slightly larger than 1.5. In all the range between $\mu = 0.01$ and $\mu = 5$ the pull-off force remains in between the rigid and JKR values (2 and 1.5 respectively). Increasing the waviness amplitude by a factor 10 ($A^{\dagger} = 1$, green squares) completely changes the picture. Figure 8 shows that there exist three distinct regimes: (i) the rigid, (ii) the transition and (iii) the JKR regime. The pull-off force remains very small and equal to the rigid limit (dot-dashed line) up to $\mu \approx 0.25$, then starts to increase up to about $|\overline{W}|_{pull-off} \simeq 3.2$ for $\mu \approx 1$ and for $\mu > 1$ tends to the JKR limit (Equation (5), dashed line). By further increasing the waviness amplitude leads to smaller pull-off forces not only in the rigid limit, but also at large Tabor parameters μ. We have indicated in Figure 8 that at $\mu \simeq 5$ the JKR prediction of the pull-off force is $|\overline{W}|_{pull-off} \simeq 20$, while numerical results give $|\overline{W}|_{pull-off} \simeq 1.1$.

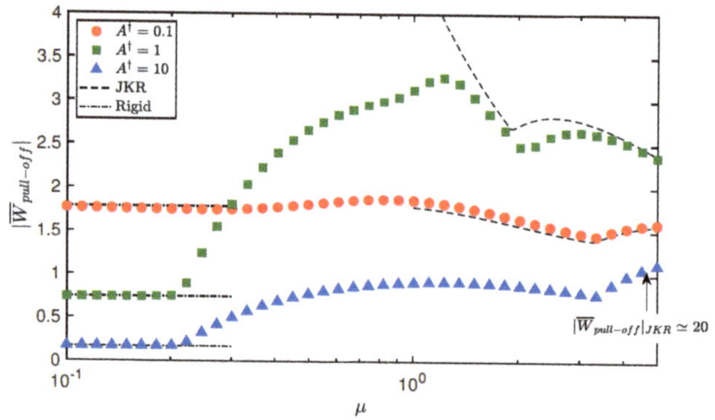

Figure 8. Pull-off force $|\overline{W}|_{pull-off}$ (absolute value) as a function of the Tabor parameter μ for $\lambda^{\dagger} = 20$, $R^{\dagger} = 100$ and $A^{\dagger} = [0.1, 1, 10]$, respectively red circles, green squares and blue triangles. Dot-dashed lines mark the rigid limit (22) whil dashed lines the Johnson–Kendall–Roberts (JKR) limit (5). For $A^{\dagger} = 10$ at $\mu = 5$ the JKR limit would give $|\overline{W}|_{pull-off} \simeq 20$.

Figure 9 shows respectively the dimensionless gap H (a) and the corresponding tractions P (b) for $\lambda^{\dagger} = 20$, $A^{\dagger} = 1$ and $\mu = [0.15, 0.67, 5]$ (respectively solid, dotted, dot-dashed line) and $A^{\dagger} = 10$, $\mu = 5$ (dashed line) at the pull-off point. Focusing on the three curves corresponding at $A^{\dagger} = 1$ one recognizes that at low Tabor parameter ($\mu = 0.15$) the maximum tensile force is reached when the sphere first touches the waviness crest, while for high Tabor parameter ($\mu = 5$, pink dot-dashed line) the typical pressure spike appears at the boundary of the contact patch. In the intermediate regime ($\mu = 0.67$) the maximum pull-off force is reached when the second crest first touches the sphere. Nevertheless, the material is too rigid to deform and the gap remains large at the first throat providing small adhesive tractions. It is useful to compare the solutions obtained for $(\mu, A^{\dagger}) = (5, 1)$ with those for $(\mu, A^{\dagger}) = (5, 10)$. In the latter case Figure 8 showed that JKR theory highly overestimates the pull-off force obtained numerically. Indeed, Figure 9 shows that the contact patch is clustered on the waviness peaks and axisymmetric grooves (internal cracks) appear, which destroy the well known enhancement mechanism of the Guduru geometry.

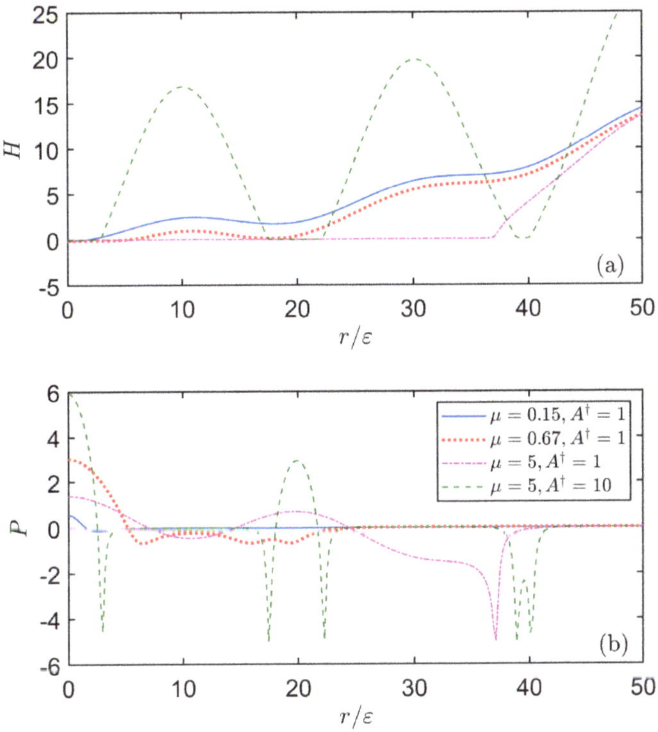

Figure 9. (**a**) Dimensionless gap H and (**b**) dimensionless tractions P versus the radial coordinate r/ε at the pull-off point for $\lambda^{\dagger} = 20$, $R^{\dagger} = 100$ and varying μ and $A^{\dagger}$ (for both panels please refer to the legend placed in panel (**b**)).

To better study the effect of the waviness amplitude $A^{\dagger}$, Figure 10 shows the dimensionless pull-off force in absolute value as a function of the ratio A/λ for $\lambda^{\dagger} = [5, 20, 30, 50]$, $R^{\dagger} = [50, 100, 200]$ and for a fixed $\mu = 3$ (see legend therein). For each value of λ Equation (5) was used to determine the pull-off force predicted by the JKR model (dashed black lines), while numerical results obtained with BEM are reported with markers (see legend in Figure 10). For amplitude to wavelength ratio below $A/\lambda \lesssim 10^{-1}$ the numerical simulations and the theoretical results are in very good agreement. For very small waviness amplitude the JKR result for the smooth sphere is obtained $\left(|\overline{W}|_{pull-off} = 1.5\right)$, while increasing A/λ adhesion enhancement takes place up to $|\overline{W}|_{pull-off} \approx 10$ for $\lambda^{\dagger} = 50$. It appears that longer wavelengths foster adhesion enhancement. For $A/\lambda \gtrsim 10^{-1}$, the pull-off force suddenly decreases and, for larger $|\overline{W}|_{pull-off}$, decays approximately with a power law, without showing a clear threshold for "stickiness" (complete elimination of adhesion), contrary to other recent theories on random roughness [40,41]. It is shown that the sphere radius markedly influences the pull-off decay, but, in the parametric region explored, it slightly affects the threshold $A/\lambda \simeq 10^{-1}$ at which the abrupt transition from adhesion enhancement to reduction takes place.

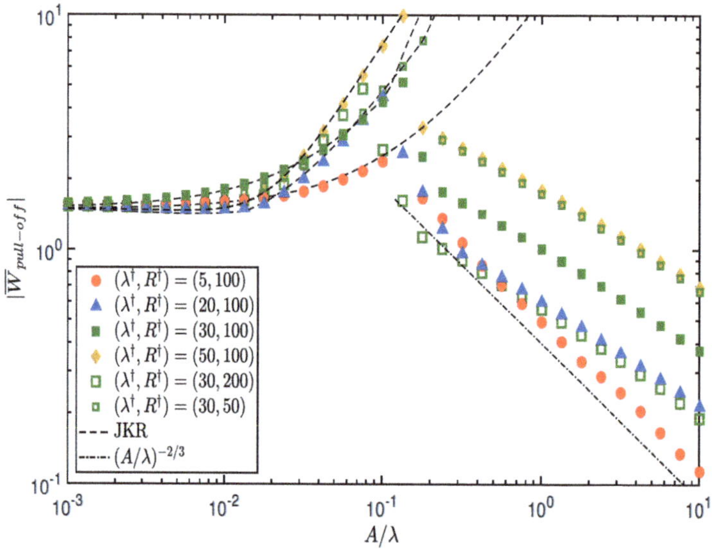

Figure 10. Dimensionless pull-off force in absolute value as a function of the ratio A/λ for the four cases $\lambda^\dagger = [5, 20, 30, 50]$, $R^\dagger = [50, 100, 200]$ and for a fixed $\mu = 3$ (see legend). Dashed lines stand for the pull-off force predicted by the JKR model (Equation (5)), markers for BEM numerical results, while the dot-dashed line is a guide to the eye.

In Figure 11 we have replotted the data in Figure 10 as effective adhesion energy $\overline{w}_{c,eff} = w_{c,eff}/w_c$ versus the Johnson parameter α_{KLJ}. Indeed, based on Kesari and Lew [21] envelope solution, Ciavarella [24] showed that in the JKR regime the effective adhesive energy at pull-off depends only on the Johnson parameter α_{KLJ}, i.e.,

$$\overline{w}_{c,eff} = \frac{2}{3}\left|\overline{W}\right|_{pull-off} = \left(1 + \frac{1}{\sqrt{\pi}\alpha_{KLJ}}\right)^2 \tag{24}$$

which is shown as a solid blue line in Figure 11. On the contrary, a competitive mechanism has been proposed by Persson and Tosatti [13], which tends to reduce the effective adhesive energy due to surface roughness in randomly rough surfaces. Persson and Tosatti [13] criterion reads

$$w_{c,eff} = w_c \frac{A_{true}}{A_{app}} - \frac{U_{el}}{A_{app}} \tag{25}$$

where A_{app} is the apparent contact area, A_{true} is the real contact area, increased due to the substrate roughness, and U_{el} is the elastic strain energy stored at full contact. The real contact area A_{true} can be written as [13]

$$A_{true} = 2\pi \int_{A_{app}} drr \left(1 + \frac{1}{2}|\nabla h|^2\right) \tag{26}$$

$$= 2\pi \int_0^{a_{app}} drr \left(1 + \frac{1}{2}\left(\frac{2\pi A}{\lambda}\right)^2 \sin^2\left(\frac{2\pi r}{\lambda}\right)\right) \tag{27}$$

where a_{app} is the apparent contact radius. Dividing Equation (27) by $A_{app} = \pi a_{app}^2$, it can be derived that for large enough a_{app}/λ

$$\frac{A_{true}}{A_{app}} \simeq 1 + \pi \left(\frac{A}{\lambda}\right)^2. \tag{28}$$

In Figure 10 we obtained the largest enhancement of the pull-off force (up to a factor 10) at about $A/\lambda \simeq 10^{-1}$, where Equation (28) would give $A_{true}/A_{app} \simeq 1.03$ (notice that $\overline{w}_{c,eff} = \frac{2}{3}\left|\overline{W}\right|_{pull-off}$), hence, in the following, we will neglect this contribution.

For a single scale waviness

$$\frac{U_{el}}{A_{app}} = \frac{E}{4(1-\nu^2)} \int d^2 q q C(q) = \frac{\pi E}{4(1-\nu^2)} \frac{A^2}{\lambda} \tag{29}$$

hence

$$\overline{w}_{c,eff} = 1 - \frac{1}{2\pi} \frac{1}{\alpha_{KLJ}^2} \tag{30}$$

which is reported as a dot-dashed red line in Figure 11. The numerical results we obtained, plotted with markers in Figure 11, show that at large α_{KLJ} the numerical results we obtained closely follow Equation (24). For smaller α_{KLJ}, instead, $\overline{w}_{c,eff}$ drops suddenly and decays by further reducing α_{KLJ} with a strong dependence on the waviness wavelength and sphere radius. Instead, the Persson–Tosatti energetic argument for adhesion reduction seems to give a lower bound to the effective work of adhesion.

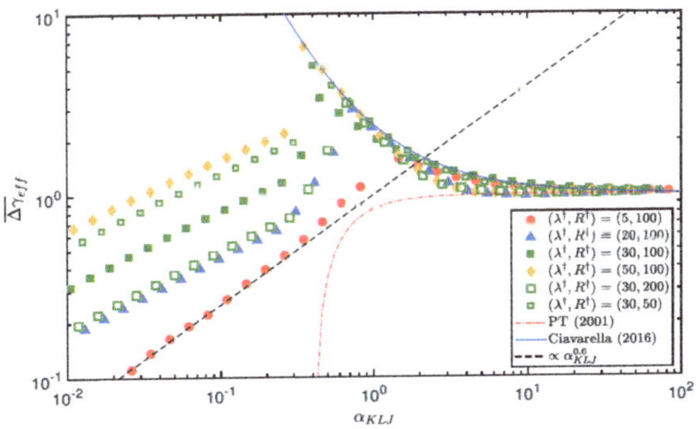

Figure 11. The data showed in Figure 10 are reported here as effective adhesion energy $\overline{w}_{c,eff} = w_{c,eff}/\Delta\gamma$ versus the α_{KLJ} for the four cases $\lambda^\dagger = [5, 20, 30, 50]$, $R^\dagger = [50, 100, 200]$ and for a fixed $\mu = 3$ (see legend). The dot-dashed line stands for the reduction criterion of Persson and Tosatti [13], the solid line for the enhancement criterion of Ciavarella [24] based on the Kesari and Lew [21] solution of the Guduru problem and the dashed line is a guide to the eye.

5.2. Hysteresis Cycle

It is well known that in adhesive contact mechanics different loading paths can be followed in loading and unloading a contact pair, which leads also to hysteretical energy dissipation. Here we show how this gets affected by the waviness amplitude A by proposing two representative examples. In Figure 12 the loading curve obtained via BEM numerical simulation is plotted as a solid red line for $\mu = 4$, $A^\dagger = 0.4$, $R^\dagger = 100$ and $\lambda^\dagger = 10$. On the same graph, the JKR loading curve for the smooth sphere (black dot-dashed curve) and for the Guduru geometry (blue dashed curve, Equation (5)) are plotted. Figure 12 shows that the numerical and the theoretical curves are very close each other and the maximum adhesive force reached is about $\left|\overline{W}\right|_{pull-off} \simeq 2$ giving a certain enhancement with respect to the smooth case. A possible loading path (in displacement control) is shown by the arrows. The jump-in and -out instability are labeled with numbers from "1" to "6" for the loading stage and with letters from "a" to "f" during unloading. Looking at Figure 12 one sees the hysteretical dissipation (proportional to the area enclosed in the hysteretical loop in Figure 12), which could be well estimated by adopting the JKR model (Equation (5)).

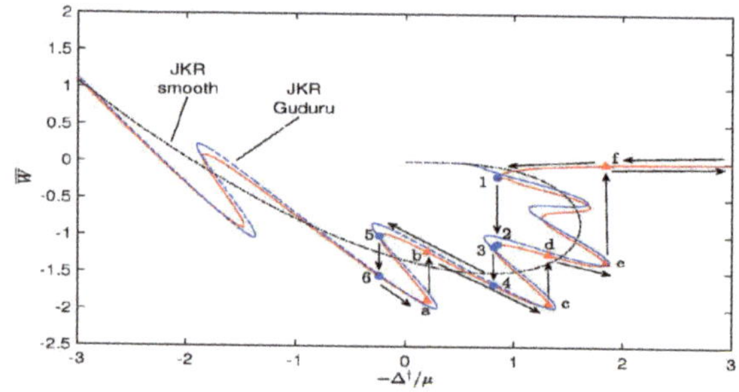

Figure 12. The dimensionless normal load $\overline{W}$ is plotted versus $-\Delta^{\dagger}/\mu$. The curve is obtained via BEM numerical simulation (solid red line) for $\mu = 4$, $A^{\dagger} = 0.4$, $R^{\dagger} = 100$ and $\lambda^{\dagger} = 10$. The JKR curve for a smooth sphere (dot-dashed black line) and for the Guduru geometry (blue dashed line, Equation (5)) are also shown. Loading and unloading paths are indicated by arrows and the jump-in and -out contact points are respectively labeled by numbers from "1" to "6" and letters from "a" to "f".

Nevertheless, the amount of dissipation is strongly influenced by the ratio A/λ and the results obtained by the JKR model (Equation (5)) may be strongly misleading. In Figure 13 the curve dimensionless normal load $\overline{W}$ versus $-\Delta^{\dagger}/\mu$ obtained numerically (red solid line) is plotted for the same parameters of Figure 12 but for $A^{\dagger} = 3$. Together with the BEM numerical results the JKR curve for the smooth sphere (black dot-dashed line) and for the Guduru geometry (blue dashed line) are shown. One immediately recognizes that the JKR model (blue dashed line) is very far from the actual loading curve (solid red curve). While the sphere approaches the wavy halfspace the JKR model predicts very large fluctuations of the normal load and relative jumps from one branch to the other that would lead to very high energy dissipation. The BEM solution, instead, gives much smaller undulations of the loading curve and smaller jumps-in and -out contact.

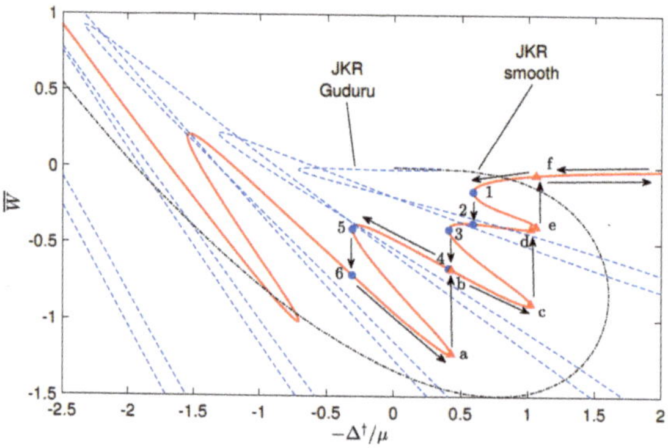

Figure 13. The dimensionless normal load $\overline{W}$ is plotted versus $-\Delta^{\dagger}/\mu$. The curve is obtained via BEM numerical simulation (solid red line) for $\mu = 4$, $A^{\dagger} = 3$, $R^{\dagger} = 100$ and $\lambda^{\dagger} = 10$. The JKR curve for a smooth sphere (dot-dashed black line) and for the Guduru geometry (Equation (5)) are also shown. Loading and unloading paths are indicated by arrows and the jump-in and -out contact points are respectively labeled by numbers from "1" to "6" and letters from "a" to "f".

5.3. Adhesion Map

To clarify the effect of A, λ and μ on the pull-off force $|\overline{W}|_{pull-off}$ we fixed the sphere radius $R^{\dagger} = 100$ and change $\mu = [10^{-1}, 5]$ and $A/\lambda = [10^{-3}, 10^{0}]$. Figures 14 and 15 shows the contour plot of the pull-off force respectively for $\lambda^{\dagger} = 20$ and $\lambda^{\dagger} = 5$. One immediately notices that larger adhesive forces are reached with longer wavelengths. Figure 14 shows that adhesion enhancement happens in a limited parameter region. For very low ratio A/λ the contact problem reduces to that of the smooth sphere on a smooth halfspace, hence by changing the Tabor parameter from $\mu = 0.01$ to $\mu = 5$ one moves from the Bradley $|\overline{W}|_{pull-off} = 2$ to the JKR solution $|\overline{W}|_{pull-off} = 1.5$. Increasing A/λ for small Tabor parameter ($\mu < 10^{-0.6} \approx 0.25$) leads to a strong reduction of the pull-off force, as indeed we are in the range where the rigid solution of the Guduru problem holds (cfr. Section 2, Figure 7). Notice that keeping λ constant and increasing A/λ leads to both increasing of $A^{\dagger}$ and α in Figure 7 heading to very strong reduction of the macroscopic pull-off force. Instead, if A/λ is increased at large Tabor parameter ($\mu \gtrsim 0.25$ for $\lambda = 20$), adhesion enhancement takes place and high pull-off forces can be reached (in Figure 14 up to $|\overline{W}|_{pull-off} \simeq 4$ for $\mu \simeq 5$). Contrary to JKR theory predictions, further increasing of the amplitude to wavelength ratio A/λ does not lead to stronger adhesive forces, but adhesion is destroyed by roughness.

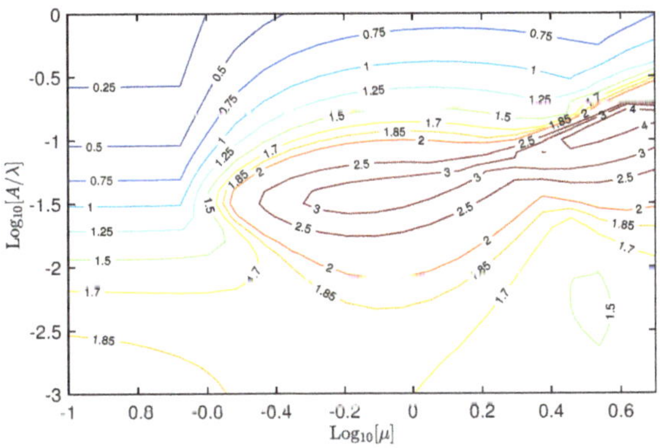

Figure 14. Contour plot of the dimensionless pull-off force (absolute value) as a function of $\mu = [10^{-1}, 5]$ and $A/\lambda = [10^{-3}, 10^{0}]$ for $\lambda^{\dagger} = 20$.

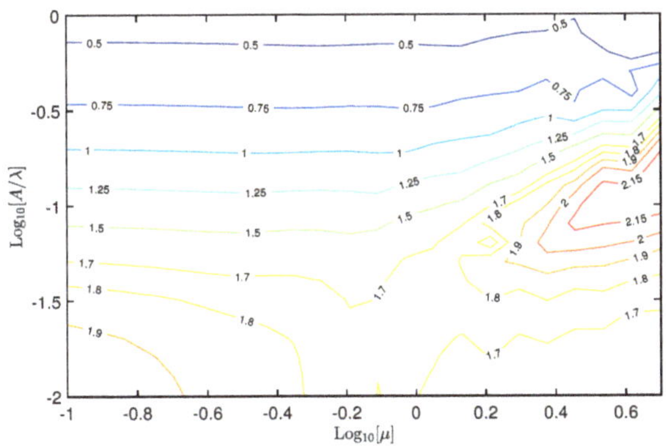

Figure 15. Contour plot of the dimensionless pull-off force (absolute value) as a function of $\mu = [10^{-1}, 5]$ and $A/\lambda = [10^{-2}, 10^{0}]$ for $\lambda^{\dagger} = 5$.

6. Conclusions

In this paper we have reconsidered the Guduru adhesive contact problem. The rigid solution has been derived, which has been shown to depend only on two parameters: the dimensionless waviness amplitude $A^{\dagger}$ and the dimensionless parameter $\alpha = AR/\lambda^2$. It has been shown that increasing $A^{\dagger}$ and α reduces the macroscopic pull-off force by orders of magnitude due to the effect of roughness. Secondly, by using a BEM numerical code with Lennard–Jones interaction law, we have investigated the effects of the waviness wavelength, amplitude and of the sphere Tabor parameter on the adhesion enhancement. It has been shown that adhesion enhancement is limited to a certain region of the plane A/λ versus μ. In particular, at low Tabor parameter increasing the ratio A/λ tends to destroy adhesion. For large Tabor parameters increasing the ratio A/λ first increases adhesion due to the Guduru enhancement mechanism, but later, for A/λ greater than about 10^{-1}, the waviness amplitude gets too large, internal cracks appear and macroscopic adhesion reduces strongly. We have shown that in this region using the JKR model to estimate both the pull-off force and the dissipated energy by hysteresis leads to very large errors as the hypothesis of compact contact area does not hold.

The enhancement effect is well captured by the Johnson parameter as derived by Ciavarella–Kesari–Lew [21,24], and is much larger than the Persson–Tosatti enhancement [13] in terms of increase of real contact area due to roughness. The Persson–Tosatti energetic argument for adhesion reduction seems to give a lower bound to the effective work of adhesion.

The axisymmetric waviness in the Guduru contact problem is highly idealized with respect to more common randomly fractal roughness, hence it is difficult to give reasonable estimates of the parameters we have introduced in our model for a fractal randomly rough surface. The analysis made is intended to shed light into the problem of adhesion enhancement with a potential application to the development of nano- and micro-mechanical systems and of bioinspired adhesives. Experimental measurements have been reported by Santos et al. [42], which show how echinoderms' tube feet exploit adhesion enhancement to increase the interfacial toughness on rough substrates. Santos et al. [42] tried to explain the interfacial toughening accounting for an increased contact area obtained when the echinoderm feet conforms to the rough substrate. We have found that adhesion enhancement may be obtained also when the latter effect is negligible.

When rough surfaces are idealized by spherical caps, a very small radius of curvature is expected at the finest scale, which suggests asperity contact takes place at very low Tabor parameters, hence adhesion enhancement seems to be very unlikely. At present, the only viable route to adhesion enhancement seems to be the design of an ad-hoc macroscopic roughness profile.

Author Contributions: Conceptualization, A.P. and M.C.; Methodology, Validation, Investigation, Writing—original draft preparation, A.P.; writing—review and editing, A.P. and M.C.; Supervision: M.C.; All authors have read and agreed to the published version of the manuscript.

Funding: This research received no external funding.

Acknowledgments: A.P. and M.C. acknowledge the support by the Italian Ministry of Education, University and Research under the Programme Department of Excellence Legge 232/2016 (Grant No. CUP-D94I18000260001). A.P. is thankful to the DFG (German Research Foundation) for funding the project PA 3303/1-1. A.P. acknowledges support from "PON Ricerca e Innovazione 2014-2020-Azione I.2" - D.D. n. 407, 27/02/2018, bando AIM (Grant No. AIM1895471).

Conflicts of Interest: The authors declare no conflict of interest.

Appendix A. BEM Formulation with Constant Pressure Discrete Elements

Equation (10) constitutes the nonlinear problem to be solved. A problem arises in evaluating the integral (11) as the kernel function $G(r,s)$ is singular in $s = r$. The common approach is to discretize Equation (11) assuming that the pressure $\sigma(s)$ has a simple form over a discrete element. To this end the simplest approach is to assume that the pressure is constant over each element. For a constant pressure $\overline{p}$ acting over the ring $c_1 < r < c_2$ the deflection at r of a single half-space is

$$u_z(r) = \frac{4\overline{p}}{\pi E^*}\left[F(c_2, r) - F(c_1, r)\right] \tag{A1}$$

where from Johnson [29]

$$F(c,r) = \begin{cases} cE\left(\frac{r}{c}\right), & r \leq c \\ r\left[E\left(\frac{c}{r}\right) - \left(1 - \left(\frac{c}{r}\right)^2\right)K\left(\frac{c}{r}\right)\right], & r > c \end{cases} \tag{A2}$$

being $K(k)$ and $E(k)$ respectively the complete elliptic integrals of first and second kind with modulus k.

Assume we have discretized the surface in N elements, so that we have $M = N + 1$ discretization points. The deflection at point r_i due to a constant pressure $\overline{p}_j$ ring in between the radii r_j and r_{j+1} is

$$u_z(r_i) = \frac{4\overline{p}_j}{\pi E^*}\left[F\left(r_{j+1}, r_i\right) - F\left(r_j, r_i\right)\right] = \frac{1}{E^*}G_{ij}\overline{p}_j \tag{A3}$$

$$G_{ij} = \frac{4}{\pi}\left[F\left(r_{j+1}, r_i\right) - F\left(r_j, r_i\right)\right] \tag{A4}$$

where the term G_{ij} within square brackets depends only on the nodal coordinates, hence by varying $i, j = 1, ..., M$ all the terms can be computed once for all.

References

1. Ciavarella, M.; Joe, J.; Papangelo, A.; Barber, J.R. The role of adhesion in contact mechanics. *J. R. Soc. Interface* **2019**, *16*, 20180738. [CrossRef] [PubMed]
2. Tadmor, R.; Das, R.; Gulec, S.; Liu, J.; N'guessan, H.; Shah, M.; Wasnik, P.; Yadav, S.B. Solid–liquid work of adhesion. *Langmuir* **2017**, *33*, 3594–3600. [CrossRef] [PubMed]
3. Vakis, A.I.; Yastrebov, V.A.; Scheibert, J.; Nicola, L.; Dini, D.; Minfray, C.; Molinari, J.F. Modeling and simulation in tribology across scales: An overview. *Tribol. Int.* **2018**, *125*, 169–199. [CrossRef]
4. Gorb, S.; Varenberg, M.; Peressadko, A.; Tuma, J. Biomimetic mushroom-shaped fibrillar adhesive microstructure. *J. R. Soc. Interface* 2007, *4*, 271–275. [CrossRef] [PubMed]
5. Skelton, S.; Bostwick, M.; O'Connor, K.; Konst, S.; Casey, S.; Lee, B.P. Biomimetic adhesive containing nanocomposite hydrogel with enhanced materials properties. *Soft Matter* **2013**, *9*, 3825–3833. [CrossRef]
6. Tang, J.; Li, J.; Vlassak, J.J.; Suo, Z. Adhesion between highly stretchable materials. *Soft Matter* **2016**, *12*, 1093–1099. [CrossRef]
7. Murphy, M.P.; Aksak, B.; Sitti, M. Gecko-inspired directional and controllable adhesion. *Small* **2009**, *5*, 170–175. [CrossRef]
8. Papangelo, A.; Lovino, R.; Ciavarella, M. Electroadhesive sphere-flat contact problem: A comparison between DMT and full iterative finite element solutions. *Tribol. Int.* **2020**. [CrossRef]
9. Sahli, R.; Pallares, G.; Papangelo, A.; Ciavarella, M.; Ducottet, C.; Ponthus, N.; Scheibert, J. Shear-induced anisotropy in rough elastomer contact. *Phys. Rev. Lett.* **2019**, *122*, 214301. [CrossRef]
10. Papangelo, A.; Scheibert, J.; Sahli, R.; Pallares, G.; Ciavarella, M. Shear-induced contact area anisotropy explained by a fracture mechanics model. *Phys. Rev. E* **2019**, *99*, 053005. [CrossRef]
11. Papangelo, A.; Ciavarella, M. On mixed-mode fracture mechanics models for contact area reduction under shear load in soft materials. *J. Mech. Phys. Solids* **2019**, *124*, 159–171. [CrossRef]
12. Ciavarella, M.; Papangelo, A. On the Degree of Irreversibility of Friction in Sheared Soft Adhesive Contacts. *Tribol. Lett.* **2020**, *68*, 81. [CrossRef]
13. Persson, B.N.J.; Tosatti, E. The effect of surface roughness on the adhesion of elastic solids. *J. Chem. Phys.* **2001**, *115*, 5597–5610. [CrossRef]

14. Violano, G.; Afferrante, L.; Papangelo, A.; Ciavarella, M. On stickiness of multiscale randomly rough surfaces. *J. Adhes.* **2019**, 1–19. [CrossRef]
15. Fuller, K.N.G.; Tabor, D. The effect of surface roughness on the adhesion of elastic solids. *Proc. R. Soc. Lond. A Math. Phys. Sci.* **1975**, *345*, 327–342.
16. Briggs, G.A.D.; Briscoe, B.J. The effect of surface topography on the adhesion of elastic solids. *J. Phys. D Appl. Phys.* **1977**, *10*, 2453. [CrossRef]
17. Guduru, P.R. Detachment of a rigid solid from an elastic wavy surface: Theory. *J. Mech. Phys. Solids* **2007**, *55*, 445–472. [CrossRef]
18. Johnson, K.L.; Kendall, K.; Roberts, A. Surface energy and the contact of elastic solids. *Proc. R. Soc. Lond. A Math. Phys. Sci.* **1971**, *324*, 301–313.
19. Guduru, P.R.; Bull, C. Detachment of a rigid solid from an elastic wavy surface: Experiments. *J. Mech. Phys. Solids* **2007**, *55*, 473–488. [CrossRef]
20. Kesari, H.; Doll, J.C.; Pruitt, B.L.; Cai, W.; Lew, A.J. Role of surface roughness in hysteresis during adhesive elastic contact. *Philos. Philos. Mag. Lett.* **2010**, *90*, 891–902. [CrossRef]
21. Kesari, H.; Lew, A.J. Effective macroscopic adhesive contact behavior induced by small surface roughness. *J. Mech. Phys. Solids* **2011**, *59*, 2488–2510. [CrossRef]
22. Waters, J.F.; Lee, S.; Guduru, P.R. Mechanics of axisymmetric wavy surface adhesion: JKR–DMT transition solution. *Int. J. Solids Struct.* **2009**, *46*, 1033–1042. [CrossRef]
23. Bradley, R.S. The cohesive force between solid surfaces and the surface energy of solids. *Lond. Edinb. Dublin Philos. J. Sci.* **1932**, *13*, 853–862. [CrossRef]
24. Ciavarella, M. On roughness-induced adhesion enhancement. *J. Strain Anal. Eng. Des.* **2016**, *51*, 473–481. [CrossRef]
25. Rumpf, H. *Particle Technology*; Chapman & Hall: London, UK, 1990.
26. Rabinovich, Y.I.; Adler, J.J.; Ata, A.; Singh, R.K.; Moudgil, B.M. Adhesion between nanoscale rough surfaces: I. Role of asperity geometry. *J. Colloid Interface Sci.* **2000**, *232*, 10–16. [CrossRef]
27. Rabinovich, Y.I.; Adler, J.J.; Ata, A.; Singh, R.K.; Moudgil, B.M. Adhesion between nanoscale rough surfaces: II. Measurement and comparison with theory. *J. Colloid Interface Sci.* **2000**, *232*, 17–24. [CrossRef]
28. Ciavarella, M. An upper bound to multiscale roughness-induced adhesion enhancement. *Tribol. Int.* **2016**, *102*, 99–102. [CrossRef]
29. Johnson, K.L. *Contact Mechanics*; Cambridge University Press: Cambridge, UK, 1985.
30. Johnson, K.L. The adhesion of two elastic bodies with slightly wavy surfaces. *Int. J. Solids Struct.* **1995**, *32*, 423–430. [CrossRef]
31. Wu, J.J. Numerical Analyses on the Adhesive Contact between a Sphere and a Longitudinal Wavy Surface. *J. Adhes.* **2014**, *91*, 381–408. [CrossRef]
32. Medina, S.; Dini, D. A numerical model for the deterministic analysis of adhesive rough contacts down to the nano-scale. *Int. J. Solids Struct.* **2014**, *51*, 2620–2632. [CrossRef]
33. Rey, V.; Anciaux, G.; Molinari, J.F. Normal adhesive contact on rough surfaces: Efficient algorithm for FFT-based BEM resolution. *Comput. Mech.* **2017**, *60*, 69–81. [CrossRef]
34. Li, Q.; Pohrt, R.; Popov, V.L. Adhesive Strength of Contacts of Rough Spheres. *Front. Mech. Eng.* **2019**, *5*, 7. [CrossRef]
35. Greenwood, J.A. Adhesion of elastic spheres. *Proc. R. Soc. Lond. Ser. A Math. Phys. Eng. Sci.* **1997**, *453*, 1277–1297. [CrossRef]
36. Feng, J.Q. Contact behavior of spherical elastic particles: A computational study of particle adhesion and deformations. *Colloids Surfaces Physicochem. Eng. Asp.* **2000**, *172*, 175–198. [CrossRef]
37. Attard, P.; Parker, J.L. Deformation and adhesion of elastic bodies in contact. *Phys. Rev. A* **1992**, *46*, 7959. [CrossRef]
38. Nayfeh, A.H.; Balachandran, B. *Applied Nonlinear Dynamics: Analytical, Computational, and Experimental Methods*; John Wiley & Sons: Hoboken, NJ, USA, 2008.
39. Maugis, D. Adhesion of spheres: the JKR-DMT transition using a Dugdale model. *J. Colloid Interface Sci.* **1992**, *150*, 243–269. [CrossRef]
40. Pastewka, L.; Robbins, M.O. Contact between rough surfaces and a criterion for macroscopic adhesion. *Proc. Natl. Acad. Sci. USA* **2014**, *111*, 3298–3303. [CrossRef]

41. Ciavarella, M. Universal features in "stickiness" criteria for soft adhesion with rough surfaces. *Tribol. Int.* **2020**, *146*, 106031. [CrossRef]
42. Santos, R.; Gorb, S.; Jamar, V.; Flammang, P. Adhesion of echinoderm tube feet to rough surfaces. *J. Exp. Biol.* **2005**, *208*, 2555–2567. [CrossRef]

lubricants

MDPI

Article

The Basin Stability of Bi-Stable Friction-Excited Oscillators

Merten Stender [1,*], Norbert Hoffmann [1,2] and Antonio Papangelo [1,3]

1 Hamburg University of Technology, 21073 Hamburg, Germany; norbert.hoffmann@tuhh.de (N.H.); antonio.papangelo@poliba.it (A.P.)
2 Imperial College London, London SW7 1AL, UK
3 Politecnico di Bari, 70125 Bari, Italy
* Correspondence: m.stender@tuhh.de

Received: 5 November 2020; Accepted: 27 November 2020; Published: 8 December 2020

Abstract: Stability considerations play a central role in structural dynamics to determine states that are robust against perturbations during the operation. Linear stability concepts, such as the complex eigenvalue analysis, constitute the core of analysis approaches in engineering reality. However, most stability concepts are limited to local perturbations, i.e., they can only measure a state's stability against small perturbations. Recently, the concept of *basin stability* was proposed as a global stability concept for multi-stable systems. As multi-stability is a well-known property of a range of nonlinear dynamical systems, this work studies the basin stability of bi-stable mechanical oscillators that are affected and self-excited by dry friction. The results indicate how the basin stability complements the classical binary stability concepts for quantifying how stable a state is given a set of permissible perturbations.

Keywords: nonlinear dynamics; basin of attraction; self-excitation; bi-stability; multi-stability

1. Introduction

The dynamics of systems affected by friction are most often studied in the context of friction-excited vibrations (FIV). Prominent examples for FIV in mechanical structures and machines range from brake systems [1–4], clutches [5], drill strings [6] to artificial hip joints [7] and others. FIV often arise through positive energy feedback from a friction interface with the structure, i.e., through self-excitation [8–10]. Sub-critical Hopf bifurcations [11,12] and isolated solution branches [13–15] are a common observation in those systems, such that bi- and multi-stable systems have been reported numerously [14,16,17]. The computation of those nonlinear responses (periodic, quasi-period orbits, chaotic trajectories) is a well-established field of research [18–21], mostly resulting in the identification of complicated bifurcation diagrams [11,13,22–24]. The stability of the solutions is usually assessed by local Lyapunov-type stability metrics [25,26]. Hence, the stability statement is often a binary one that measures the state's robustness against *small* perturbations. However, the actual size of *permissible perturbations*, i.e., those for which the trajectory would still return back to the state, is not given. In a multi-stable system configuration, the long term steady-state behavior thus depends on the choice of initial conditions or the size of instantaneous perturbations. Once the system enters another basin of attraction, severe jumping phenomena may occur. Typically, such jumps are related to a change from a stable steady sliding state to high-intensity periodic vibrations or stick-slip cycles [27–29], or from one periodic solution to another periodic solution [11].

This work investigates a rather novel technique denoted as *basin stability* to estimate the size of the system's basins of attraction in a subset of the state space. The basins' size estimation can then be considered a global stability metric, i.e., indicating how likely the system is to end up on one of the co-existing stable steady-states. Therefore, those probabilities add new insights to the rather binary stability statements derived from local perturbation-based approaches. We study a friction oscillator excited by a falling friction slope and a second oscillator excited through binary flutter instability. Our results indicate that the basin stability analysis is a robust and easily applicable model-agnostic technique. It can reveal the *actual* picture of the long-term behavior for a given set of perturbations, thus augmenting classical bifurcation and stability studies. Using the basin stability analysis, some solutions can even be ruled out if one can guarantee strict control over the instantaneous perturbations to system trajectories or operating conditions.

2. The Concept of Basin Stability

We study nonlinear dynamic systems

$$\dot{\mathbf{x}} = \mathbf{f}(\mathbf{x}, t), \quad \mathbf{x} \in \mathbb{R}^N \tag{1}$$

with the states $\mathbf{x}(t)$ in the N-dimensional state space. The long-term asymptotic behavior is denoted as *attractor* $\mathcal{A}$ [30] throughout this work. Typically, the Lyapunov spectrum $\Lambda = [\lambda_1, \ldots, \lambda_N]$ is assessed to characterize the linear stability of a state $\mathbf{x}$ against small perturbations. For fixed points, the Lyapunov exponents are equivalent to the system's eigenvalues derived from the complex eigenvalue analysis (CEA). The real parts of the eigenvalues indicate linear stability to a *small* perturbation about the fixed point. The sizes of the real parts indicate the strength of attraction ($\lambda < 0$) or rejection ($\lambda > 0$) for stable or unstable directions in state space, respectively. However, the eigenvalues do not encode a piece of information about the permissible size of perturbations that are still attracted by the fixed point. While this is not an issue for systems that feature only a single stable solution, the situation is different for systems featuring multiple stable solutions. For these systems, local stability concepts may have only a limited validity: a non-small perturbation of a state can result in a jump to another attractor. Hence, global stability concepts are required to assess the size of permissible perturbations, i.e., to characterize the basins of attraction for all solutions. The basin of attraction

$$\mathcal{B}(\mathcal{A}) = \left\{ \mathbf{x}_0 \in \mathbb{R}^N \mid \lim_{t \to \infty} \mathbf{x}(t) = \mathcal{A}, \quad \mathbf{x}_0 = \mathbf{x}(t=0) \right\} \tag{2}$$

denotes the subset of states that converge to the same attracting set $\mathcal{A}$. The basin boundaries are related to unstable solutions of the system which represent separatrices of the basins in state space. Depending on the size and shape of its basin, an attractor can be more or less robust against non-small perturbations. There are multiple ways to compute the basins of attraction, e.g., through Lyapunov functions [31]. These methods are known for some canonical, low-dimensional, and well-studied systems. However, they are not readily available, or straight-forward to compute, for any generic and high-dimensional nonlinear dynamical system, such as frictional oscillators which are studied in this work.

The basin stability proposed by Menck et al. [32,33] is a global stability concept for complex systems that aims at measuring stability against non-small perturbations by a volume-based probabilistic approach. Conceptually, the basin stability measures the volumetric share of all basins of attraction in a hypervolume of the state space. For a computationally feasible solution, a distribution $\rho(\mathbf{x})$ of perturbations is drawn from a reference subset $\mathcal{Q} \subset \mathbb{R}^N$ of the state space, representing a set of states to which the system may be pushed to through non-small perturbations with $\int_{\mathcal{Q}} \rho(\mathbf{x})\, d\mathbf{x} = 1$. For each perturbation, the steady-state behavior of the dynamical system is obtained through time-marching integrations. Then, the fraction of perturbed states that converged to the specific attractor $\mathcal{A}$ denotes an estimate for the basin stability $\mathcal{S}_{\mathcal{B}}(\mathcal{A})$, i.e., [32,34]

$$\mathcal{S}_{\mathcal{B}}(\mathcal{A}) = \int \kappa_{\mathcal{B}}(\mathcal{A})(\mathbf{x})\, \rho(\mathbf{x})\, \mathrm{d}\mathbf{x}, \quad \kappa_{\mathcal{B}}(\mathcal{A})(\mathbf{x}) = \begin{cases} 1, & \text{if } \mathbf{x} \in \mathcal{B}(\mathcal{A}) \\ 0, & \text{otherwise} \end{cases}. \tag{3}$$

Here, $\kappa_{\mathcal{B}}$ denotes an indicator function that classifies a steady state solution $\mathbf{x}(t)$ to belong to the attractor $\mathcal{A}$. Therefore, $\mathcal{S}_{\mathcal{B}}(\mathcal{A})$ is an estimate for the volume share of the basin of attraction $\mathcal{B}_{\mathcal{A}}$ given the reference subset $\mathcal{Q} \subset \mathbb{R}^N$ sampled by $\rho(\mathbf{x})$ [32,35]. Naturally, for a k-multi-stable system, the basin stability values of all k attractors add up to unity $\sum_i^k \mathcal{S}_{\mathcal{B}}(\mathcal{A}_i) = 1$. The size, i.e., the number of states, of the dynamical system to be studied by the basin stability is only limited by computational power for the Monte Carlo simulations. As the basin stability computation can be considered a repeated Bernoulli experiment [32], the standard error of the basin stability estimate is

$$e = \frac{\sqrt{\mathcal{S}_{\mathcal{B}}(1 - \mathcal{S}_{\mathcal{B}})}}{\sqrt{n}} \tag{4}$$

which can be used to find a subset $\mathcal{Q}$ that ensures a low standard error. Recently, systems with fractal, riddled, and intermingled basin boundaries were studied [33] indicating the robustness of the basin stability concept. All basin stability computations in this work were obtained from the open-source package `bSTAB` [36] available at https://github.com/TUHH-DYN/bSTAB/tree/v1.0.

Figure 1 displays a schematic for illustrating the practical computation of basin stability values. A nonlinear dynamical system with two states $\mathbf{x} = [x_1, x_2]$ is studied (In fact, the system is the single-degree-of-freedom frictional oscillator to be discussed in Section 3). The system exhibits three solutions: A stable equilibrium position ($\mathbf{x}_{\text{EP}}$), an unstable periodic orbit ($\mathbf{x}_{\text{UPO}}$), and a stable limit cycle ($\mathbf{x}_{\text{LC}}$). The distribution of perturbations $\rho(\mathbf{x})$ is chosen such that all solutions are contained in $\mathcal{Q}$ and $n = 100$ samples are drawn uniformly at random. The trajectories starting from $n_{\text{EP}} = 37$ states in the basin $\mathcal{B}_{\text{EP}}$ converge towards the equilibrium position, while $n_{\text{LC}} = 63$ states are located in the basin $\mathcal{B}_{\text{LC}}$ and thus converge to the stable limit cycle. As a result, the basin stability estimates are $\mathcal{S}_{\mathcal{B}}(\text{EP}) = 0.37$ and $\mathcal{S}_{\mathcal{B}}(\text{LC}) = 0.63$, respectively. Because the separatrix, which is the unstable periodic orbit, is explicitly known for the system, the basin volumes can be determined analytically. The exact volumetric fractions of $\mathcal{B}_{\text{EP}}$ and $\mathcal{B}_{\text{LC}}$ in $\mathcal{Q}$ are 0.3275 and 0.6725, respectively. Therefore, the basin stability computed from $n = 100$ samples is a good approximation for the system at hand (Appendix A.2 indicates that $n \approx 300$ samples are required for a very close approximation of the analytical results).

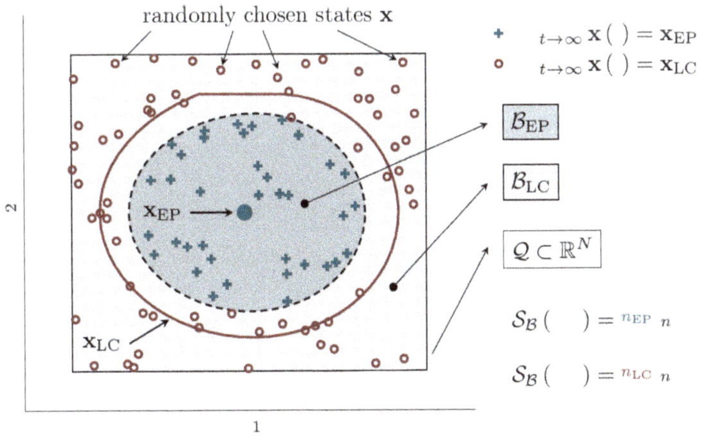

Figure 1. Schematic of the basin stability calculation. In the two-dimensional state space, two stable attractors EP (equilibrium position) and LC (limit cycle) co-exist. The respective basins of attraction $\mathcal{B}_{\text{EP}}$ and $\mathcal{B}_{\text{LC}}$ are separated by an unstable periodic orbit (indicated by the dashed line). The steady-state behaviors of $n = 100$ randomly sampled states are used to estimate the volume shares of the basins of attraction in the subset $\mathcal{Q}$. The resulting basin stability estimates are $\mathcal{S}_{\mathcal{B}}(\text{EP}) = 0.37$, $\mathcal{S}_{\mathcal{B}}(\text{LC}) = 0.63$ for this example.

3. Bi-Stable Oscillator with Falling Friction Slope

As a first system, we study the dynamics and the stability of a single-degree-of-freedom oscillator $m\ddot{x} + c\dot{x} + kx = F$, see Figure 2a, with velocity-dependent friction as proposed by Papangelo et al. [12]. Specifically, the friction characteristic $\mu\,(v_{\mathrm{rel}})$ is a velocity-dependent weakening function

$$
\begin{aligned}
v_{\mathrm{rel}} \neq 0: &\quad F = -N\mu\,(v_{\mathrm{rel}})\,\mathrm{sign}\,(v_{\mathrm{rel}})\,, \quad v_{\mathrm{rel}} = \dot{x} - v_{\mathrm{d}} \\
v_{\mathrm{rel}} = 0: &\quad |F| < \mu_{\mathrm{st}} N \\
&\mu(v_{\mathrm{rel}}) = \mu_{\mathrm{d}} + (\mu_{\mathrm{st}} - \mu_{\mathrm{d}})\exp\left(-\frac{|v_{\mathrm{rel}}|}{v_0}\right)
\end{aligned}
\tag{5}
$$

featuring the static friction coefficient $\mu\,(0) = \mu_{\mathrm{st}}$, the dynamic friction coefficient $\mu\,(v_{\mathrm{rel}} \to +\infty) = \mu_{\mathrm{d}}$, the reference velocity v_0 and the contact normal load N. The non-dimensional form of the equations of motion is obtained through normalization $(\tilde{\cdot})$ of the quantities accordingly to the work of Papangelo et al. [12]. The velocity-dependence introduces a dynamic instability that gives rise to friction-induced vibrations (FIV) for $0 \leq \tilde{v}_{\mathrm{d}} \leq 1.84$. Moreover, the friction nonlinearity enables the system to exhibit a bi-stable behavior, such that a stable steady sliding state and a stable stick-slip cycle co-exist for a range of belt velocities $1.11 \leq \tilde{v}_{\mathrm{d}} \leq 1.84$, see Figure 2b. At $\tilde{v}_{\mathrm{d}} = 1.15$, the steady sliding state loses stability at a subcritical Hopf bifurcation point. In the bi-stability regime, and depending on the initial condition or instantaneous perturbations, the system will either end up in the low-energy steady sliding state, or on the high-intensity stick-slip cycle. Both solutions are locally stable and attractive, i.e., robust against small perturbations.

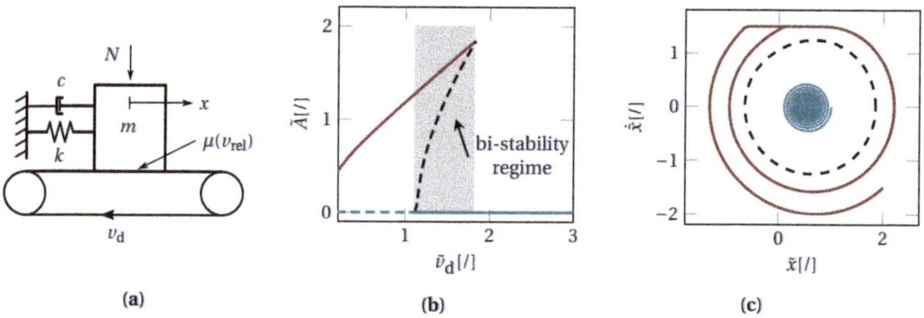

Figure 2. (**a**) single-degree-of-freedom frictional oscillator, (**b**) bifurcation diagram for the non-dimensional belt velocity $\tilde{v}_{\mathrm{d}}$, and (**c**) phase plane for $\tilde{v}_{\mathrm{d}} = 1.5$. Stable (unstable) solutions are indicated by solid (dashed) lines. The stable steady sliding state (blue spiral trajectory) co-exists with the unstable periodic orbit (black dashed line) and the stable stick-slip limit cycle (red trajectory). The non-dimensional system $(\tilde{\cdot})$ is evaluated for $\mu_{\mathrm{d}} = 0.5$, $\mu_{\mathrm{st}} = 1.0$, $\xi = 0.005$, $N = 1.0$ and $\tilde{v}_0 = 0.5$.

For this minimal system, the basin boundaries are directly accessible through the unstable periodic orbit (UPO). However, if this knowledge was not available, the probability of arriving on one of the two steady states would be unknown. Figure 1 displays a sampling with $n = 100$ points uniform at random from $\mathcal{Q}\,(x, \dot{x}) : [-3, 3] \times [-2, 2]$ at $\tilde{v}_{\mathrm{d}} = 1.5$, and the resulting basin stability values $\mathcal{S}_{\mathcal{B}}\,(\mathrm{FP}) = 0.37$ and $\mathcal{S}_{\mathcal{B}}\,(\mathrm{LC}) = 0.63$. Hence, for this $\rho\,(\mathbf{x})$, it is more likely to arrive on the high-amplitude limit cycle solution than on the steady sliding fixed point.

To complement the bifurcation diagram and the complex eigenvalue analysis, the basin stability of the fixed point and limit cycle solution is derived along the normalized belt velocity parameter. In particular, at each velocity value $n = 1000$ initial conditions are drawn from a uniform random distribution in $\mathcal{Q}\,(x, \dot{x}) : [0.5, 2.5] \times [-2, 0]$, i.e., positive initial displacement and negative initial velocity. Figure 3 depicts the eigenvalue's real part and the basin stability. As $\tilde{v}_{\mathrm{d}}$ decreases, the real part grows until it crosses into the positive plane at $\tilde{v}_{\mathrm{d}} = 1.15$. This rather smooth behavior nicely indicates the transition into linear instability of the fixed point solution. However, the eigenvalues at the exemplary points $\tilde{v}_\mathrm{d} = 1.3$ and $\tilde{v}_\mathrm{d} = 1.7$ would not allow a statement about the system's probability to converge to this solution instead of converging to the periodic orbit. Additionally, the eigenvalue obviously does not indicate the existence of the competing stable periodic solution in this parameter range. At this point the basin stability analysis comes into play: Below the Hopf bifurcation point, all trajectories converge to the periodic orbit, hence $\mathcal{S}_\mathcal{B}\,(\mathrm{LC}) = 1.0$, and above the bi-stability regime all trajectories converge to the globally stable fixed point, i.e., $\mathcal{S}_\mathcal{B}\,(\mathrm{FP}) = 1.0$ for $\tilde{v}_\mathrm{d} > 1.84$. For the chosen subset $\mathcal{Q}$, the periodic orbit is the dominating behavior in the lower parameter range of the bi-stable regime. For increasing relative velocity the probabilities, i.e., the basin stability values, are more balanced for arriving either on the LC or the FP. For $\tilde{v}_\mathrm{d} > 1.6$ the fixed point is the more probable solution to arrive at. Therefore, the basin stability values add an important insight and complement the binary stability statements given by the eigenvalues. Using the basin stability, it is now possible to state *how* stable a solution is against arbitrary and possibly *non-small* perturbations. For more realistic systems, this statement may be of even larger value than the binary stability statement given by local metrics.

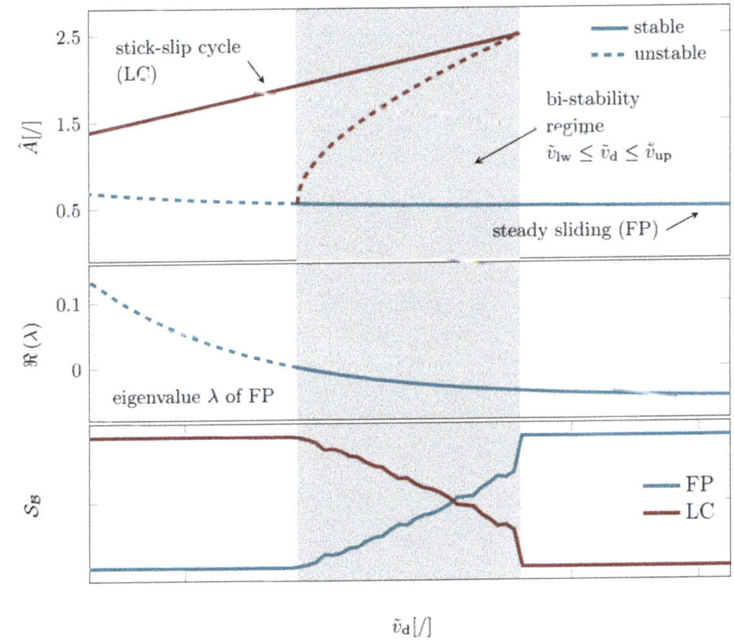

Figure 3. Bifurcation diagram (**top**), real eigenvalue (**middle**) and basin stability (**bottom**) of the single-DOF friction oscillator along the relative sliding velocity.

4. Bi-Stable Oscillator with Mode-Coupling

As a second system, we study a frictional oscillator [8,11], which (in contrast to the first system) experiences FIV through a mode-coupling instability. The system features a main oscillating mass that is in dry Coulomb-type frictional contact with a conveyer belt. A second mass is connected to the main mass through a nonlinear joint element in diagonal direction, thereby geometrically coupling the vertical and horizontal movement of the main mass. The relative sliding velocity is assumed to always be positive, such that no stick-slip cycles can arise. For the nonlinear joint element, a cubic stiffness nonlinearity k_nl is chosen [11]. The equations of motion and parameter values are given in Appendix B and the model is displayed in Figure 4.

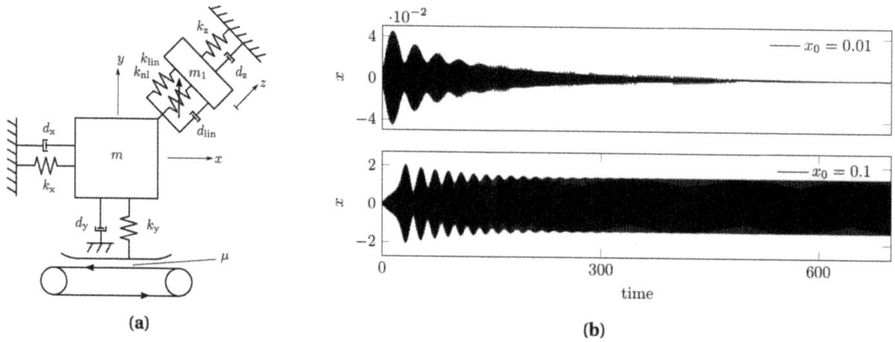

Figure 4. (**a**) Frictional oscillator with nonlinear joint and mode-coupling instability [11]. (**b**) Trajectories obtained in the reference configuration (see Appendix B) for two different initial conditions of the horizontal displacement x (all other states were kept at 0).

Previous studies have revealed the complicated bifurcation behavior of this system, including super- and sub-critical Hopf bifurcations as well as isolated solution branches [11,13,14]. In this study, a variation of the horizontal stiffness k_x is performed. A sub-critical Hopf bifurcation point is found at $k_x = 32.3$, see Figure 5a. Below, a stable limit cycle and the unstable fixed point exist. Above this value there is a bi-stable range up to $k_x = 33.0$ with a co-existing stable limit cycle and the stable fixed point. The eigenvalues' real parts in Figure 5b exhibit the classical forking behavior that is related to the mode-coupling instability mechanism in this system. At the point of instability, one eigenvalue crosses into the positive plane. The basin stability $\mathcal{S}_\mathcal{B}$ of both stable solutions is computed for $n = 500$ random initial conditions drawn from $\mathcal{Q}(x, \dot{x}) : [0, 0.5] \times [0, 0.25]$ (all other initial conditions are fixed to 0). Figure 5c depicts the basin stability as a function of the horizontal stiffness. In the bi-stability range $32.3 \leq k_x \leq 33.0$ the basin stability values indicate that the limit cycle solution is the dominating one for lower stiffness values. For larger stiffness values the fixed point solution is the most probable for our choice of $\mathcal{Q}$. Hence, within this rather short bi-stability range, a minor variation of the horizontal stiffness value would crucially affect the probability of arriving either on the low-energy steady-sliding state, or on the high-energy limit cycle, which may cause increased wear, audible vibrations and other effects in realistic systems. Such kind of statement about the global stability regarding non-small perturbations would not have been easily accessible through the bifurcation diagram or the local stability analysis.

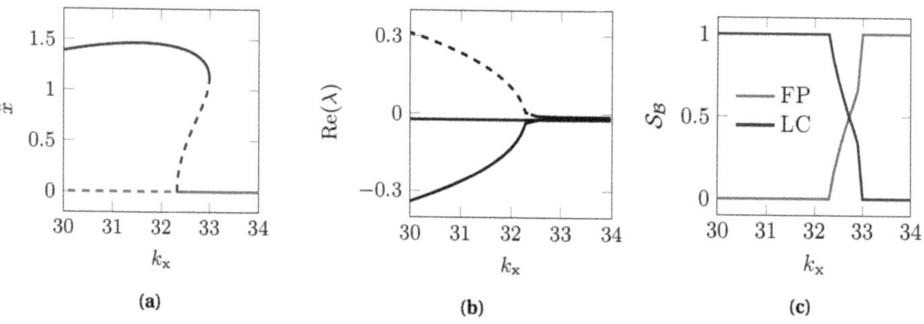

Figure 5. (**a**) bifurcation diagram for the horizontal stiffness parameter, (**b**) eigenvalues' real parts and (**c**) basin stability of the fixed point and limit cycle solution. $\hat{x}$ denotes the maximum amplitude of $x(t)$ along one vibration period. Solid and dashed lines indicate stable and unstable solutions, respectively.

These results are clearly related to the shape of the unstable periodic orbit, i.e., the separatrix of both basins of attraction. While the qualitative basin stability values for a variation in the initial conditions for x would have been readable from the bifurcation diagram in Figure 5a, this task quickly becomes complex once more degrees-of-freedom (DOFs) are considered. For example, let us consider a reference subset $\mathcal{Q}$ that captures certain variations for multiple DOFs, instead of variations for a single DOF as shown before. Figure 6 displays the basin stability values for three different choices of $\mathcal{Q}$, i.e., different variations of the range of possible initial conditions:

$$\begin{aligned} \mathcal{Q}_1\,(x,y) &: [0,0.25] \times [0,0.5] \\ \mathcal{Q}_2\,(x,y,\dot{x},\dot{y}) &: [0,0.25] \times [0,0.25] \times [-0.1,0.1] \times [-0.2,0.2] \\ \mathcal{Q}_3\,(x,y,z,\dot{x},\dot{y},\dot{z}) &: [0,0.25] \times [0,0.25] \times [0,0.25] \times [-0.1,0.1] \times [-0.1,0.1] \times [-0.1,0.1]\,. \end{aligned} \tag{6}$$

In the first case, some initial variations in the horizontal position and large variations in the vertical displacement of the main mass are allowed. In the second case, variations in the initial velocities are studied, and in the third case also variations in the secondary mass' initial conditions are considered. Such scenarios would quickly become impractical for studying permissible perturbations, i.e., the global stability of each solution, using bifurcation diagrams and subdividing the state space by the unstable solutions. The concept of basin stability automates this process through the Monte Carlo sampling, allowing for a easy-scaling and consistent estimation of the relevant basin volumes.

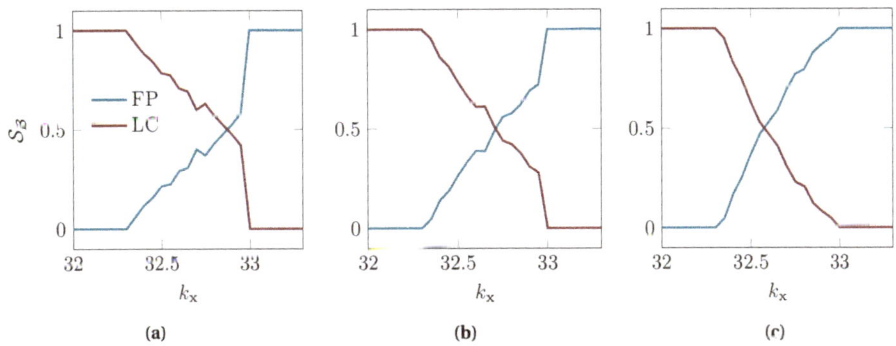

Figure 6. Basin stability values in the bi-stability range for the reference sets of initial conditions $\mathcal{Q}_1$ (**a**), $\mathcal{Q}_2$ (**b**), and $\mathcal{Q}_3$ (**c**) defined in Equation (6).

In fact, even though the three reference sets are very different in their value ranges, the resulting basin stability analysis displayed in Figure 6 does not change qualitatively. The *turning point*, i.e., the point after which the FP solution dominates over the LC solution for increasing values of k_x, changes only slightly: For $\mathcal{Q}_1$ this point is found at $k_x = 32.9$, while it is $k_x = 32.7$ and $k_x = 32.55$ for $\mathcal{Q}_2$ and $\mathcal{Q}_3$, respectively. Hence, the basin stability is not very sensitive to the choice of $\mathcal{Q}$ for this system. In a situation in which the overall qualitative behavior of the basin stability values may have seem obvious, the quantitative evaluation would have become difficult to obtain from the bifurcation diagrams. Especially for higher-dimensional systems and specific subset choices the basin stability analysis represents a highly robust approach to estimate the probability of arriving on either of the competing solutions, which we will illustrate in the next section.

5. Bi-Stable Oscillator with Isolated Periodic Solution

The third dynamical system studied in this work is a weakly damped variant of the system proposed in the previous section and sketched in Figure 4. Here, the damping parameters are reduced by a factor of 10 to $d_x = d_y = d_z = d_{\text{lin}} = 0.002$. This system configuration has already been studied in [13,14] where the authors found an isolated solution branch resulting from the damping variation. Figure 7a displays the bifurcation diagram for the horizontal stiffness k_x. The fixed point solution loses stability through a sub-critical Hopf bifurcation at $k_x = 32.24$ to a limit cycle solution, hereafter denotes as LC1. Interestingly, a second stable limit cycle solution is born for $k_x < 29.9$, which is found to be an isolated branch [14], hereafter denoted as LC2. That is, this solution is not connected to any other solution path. As a result, the system may jump from the fixed point solution to the first limit cycle for $32.24 \leq k_x \leq 33.0$, and then from the limit cycle to the isolated branch for $k_x < 29.9$. Hence, within a rather narrow parameter range, two jumping phenomena between different solutions may occur. It is, therefore, of great interest to investigate the probability of arriving on either of those solutions for some prescribed set of initial conditions.

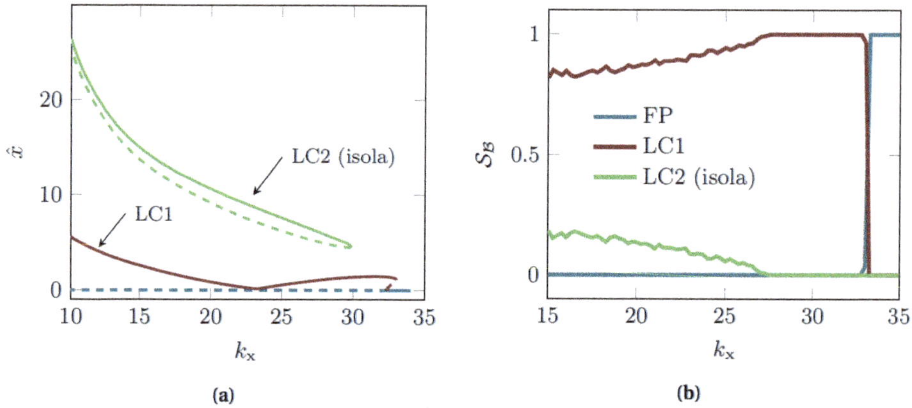

Figure 7. Bifurcation diagram for the weakly damped friction oscillator exhibiting an isolated solution branch (**a**) and basin stability values (**b**) for all three stable solutions along the horizontal stiffness k_x. Initial conditions for each solution are given in Appendix C.

Figure 7b displays the basin stability values for both periodic orbits and the fixed point solution. For the reference subset, we arbitrarily chose $\mathcal{Q}\,(x, y, z, \dot{x}, \dot{y}, \dot{z}) : [0, 10] \times [0, 10] \times [0, 10] \times [-2, 2] \times [-2, 2] \times [-2, 2]$ using $n = 1000$ sampling points. For for the bi-stability range featuring the two periodic solutions LC1 and LC2 ($k_x < 29.9$) the basin stability analysis reveals that LC1 is the by far most probable solution. A maximum of 21% of the trajectories converge to the isolated solution branch, while the remaining trajectories converge to the first periodic orbit. Particularly interesting is the parameter regime $27.4 \leq k_x \leq 29.9$. Here, the basin stability indicates that LC1 is globally stable, even though the stable isola still co-exists. However, due to the choice of $\mathcal{Q}$, no initial conditions were drawn for the basin related to LC2. Hence, if the range of initial conditions and perturbations can be quantified or limited for some specific system, the basin stability analysis can also help to rule out jumping phenomena between co-existing solutions.

Another interesting observation is the following: the basin stability values in this specific setting do not follow the qualitative trend of the respective amplitudes reported in Figure 7a. $\mathcal{S}_{\mathcal{B}}\,(\text{LC1})$ keeps increasing along the stiffness parameter, while the corresponding amplitude of the horizontal vibration amplitude shows a different behavior. Theoretically, it is clear that the vibration amplitudes do not relate to the size of the basins of attraction. However, on the first sight classical bifurcation diagrams may suggest that one solution is *more attractive* if it has a larger vibration amplitude. At this point, the basin stability represents a technique to quantify the attractiveness in a highly consistent manner.

Lastly, we discuss our previous thought on the benefits of having a robust methodical approach to estimating the basin volumes through Monte Carlo sampling irrespective of the dynamical system at hand (so-called *model-agnostic* techniques). Especially for such low-dimensional systems as shown before, one might raise the issue of using computation-heavy sampling methods, even though the basins of attraction are readily available once the bifurcation diagram is known. Figure 8 displays the state space of each DOF at $k_x = 27$, hence in a configuration where the two periodic orbits co-exists. It becomes clear that even for this 3 DOF oscillator (6 states), the analytical calculation of the basin volumes can quickly become a challenge. There is no straight-forward way to computing the volumes in the six-dimensional space from the intertwined basins separated by the unstable orbits, especially looking at the z coordinate. Therefore, the basin stability analysis is not only relevant for systems featuring larger number of states, but also for rather low-dimensional systems.

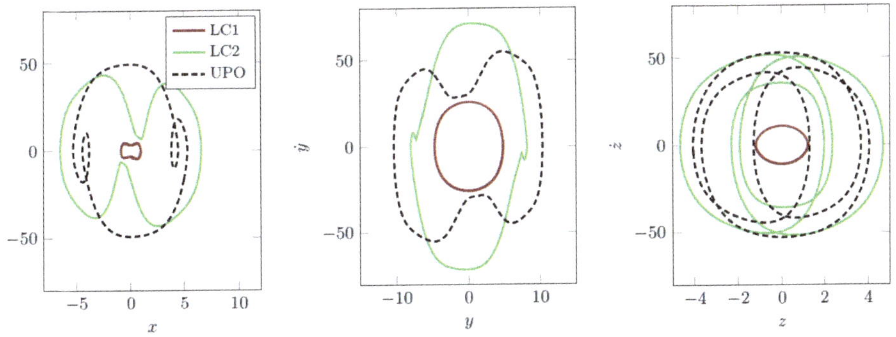

(a) (b) (c)

Figure 8. State space of all DOFs (horizontal direction in (**a**), vertical direction in (**b**) and diagonal direction in (**c**)) at $k_x = 27.0$ for the weakly damped oscillator.

6. Conclusions

This work proposed augmenting the classical local stability analysis of friction-excited oscillators by their basin stability. The concept of basin stability allows assigning global stability metrics to multi-stable solutions in a highly automated manner including error estimates. For three different friction excited systems, we show that the knowledge of global stability with respect to a specific set of initial conditions can provide important insights into the long-term dynamics. Particularly for well-controlled perturbations, this approach allows estimating the probabilities of arriving on either of multiple stable solutions, and even to rule out some steady-state behavior. As a result, we suggest to include the basin stability analysis into the toolbox of techniques that are applied to study the nonlinear dynamics of multi-stable systems, especially when operating conditions are well-known.

Author Contributions: Conceptualization, M.S. and N.H.; methodology, M.S.; software, M.S.; validation, M.S.; formal analysis, M.S.; investigation, M.S. and A.P.; resources, N.H.; data curation, M.S.; writing—original draft preparation, M.S.; writing—review and editing, N.H. and A.P.; visualization, M.S.; supervision, N.H. and A.P.; project administration, N.H.; funding acquisition, N.H. All authors have read and agreed to the published version of the manuscript.

Acknowledgments: Publishing fees supported by Funding Programme **Open Access Publishing** of Hamburg University of Technology (TUHH).

Funding: This research was funded by the German Research Foundation (Deutsche Forschungsgesellschaft DFG) within Priority Program 1897 'calm, smooth, smart', grant number 314996260.

Conflicts of Interest: The authors declare no conflict of interest. The funders had no role in the design of the study; in the collection, analyses, or interpretation of data; in the writing of the manuscript, or in the decision to publish the results.

Abbreviations

The following abbreviations are used in this manuscript:

DOF	degree-of-freedom
FIV	friction-induced vibrations
FP	fixed point
LC	limit cycle

Appendix A. Single-DOF Oscillator

Appendix A.1. Equations of Motion

Using $\omega_n = \sqrt{\frac{k}{m}}$, $\xi = \frac{c}{2\sqrt{km}}$, $x_0 = \frac{N}{k}$, and $\tau = \omega_n t$, $\frac{d}{dt} = \omega_n \frac{d}{d\tau}$ we re-write Equation (5) into

$$\ddot{\tilde{x}} + 2\xi\dot{\tilde{x}} + \tilde{x} = \tilde{F} \tag{A1}$$

where $(\tilde{\cdot})$ indicates a non-dimensional quantity.

Appendix A.2. Convergence of Basin Stability Values

The number of samples n is varied to answer the question of how many samples from $\mathcal{Q}$ are required for a robust approximation of the basin stability values. Figure A1 displays the convergence of the basin stability values for the single-DOF oscillator case and the corresponding analytical values.

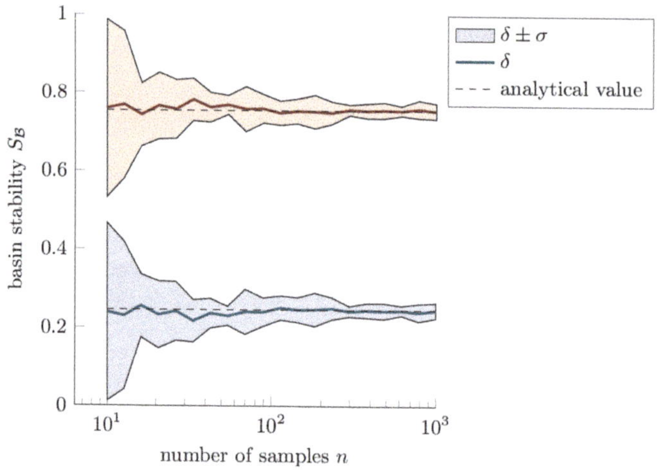

Figure A1. Effect of increasing the number of samples for estimating the basin stability values at $\tilde{v}_d = 1.5$. For each value n, the calculation has been repeated ten times. Mean values δ and the standard deviation σ are reported along with the analytical values.

Appendix B. Mode-Coupling Instability Oscillator

The equations of motion are given by

$$\mathbf{M}\ddot{\mathbf{x}} + (\mathbf{D} + \mathbf{G})\,\dot{\mathbf{x}} + (\mathbf{K} + \mathbf{N})\,\mathbf{x} + \mathbf{f}_{\text{nl}} = \mathbf{0}, \quad \mathbf{x} = [x, y, z]^{\top},$$

$$\mathbf{M} = \begin{bmatrix} m & 0 & 0 \\ 0 & m & 0 \\ 0 & 0 & m_1 \end{bmatrix}, \quad \mathbf{D} = \begin{bmatrix} d_x & 0 & 0 \\ 0 & d_y & 0 \\ 0 & 0 & d_x \end{bmatrix}, \quad \mathbf{G} = \mathbf{0}, \quad \mathbf{K} = \begin{bmatrix} k_x & -0.5k_y\mu & 0 \\ -0.5k_y\mu & k_y & 0 \\ 0 & 0 & k_z \end{bmatrix}, \tag{A2}$$
$$\mathbf{N} = \begin{bmatrix} 0 & -0.5k_y\mu & 0 \\ 0.5k_y\mu & 0 & 0 \\ 0 & 0 & 0 \end{bmatrix}, \quad F_{\text{nl}} = uk_{\text{lin}} + u^3 k_{\text{nl}} + \dot{u}d_{\text{lin}}, \quad \mathbf{f}_{\text{nl}} = \begin{bmatrix} -\frac{\sqrt{2}}{2} \\ -\frac{\sqrt{2}}{2} \\ 1 \end{bmatrix} F_{\text{nl}}$$

where u is the relative displacement in the joint between the main mass and the secondary mass, given by $u = -\frac{\sqrt{2}}{2}x - \frac{\sqrt{2}}{2}y + z$. The parameter values for the reference configuration are given by $m = m_1 = 1$, $k_x = 32.5$, $k_y = 20$, $k_z = 100$, $k_{\text{lin}} = 10$, $k_{\text{nl}} = 5$, $d_x = d_y = d_z = d_{\text{lin}} = 0.02$, $\mu = 0.65$.

Appendix C. Mode-Coupling Instability Oscillator with Isolated Solutions

Compared to the system configuration given in Appendix B, the damping parameter values are set to $d_x = d_y = d_z = d_{lin} = 0.002$. Initial values on the periodic orbits at $k_x = 11$ for the weakly damped system configuration read

$$\begin{aligned} \mathbf{y}_0 &= \begin{bmatrix} -1.1366 & -4.5527 & -1.2077 & -0.0125 & 0.0722 & -0.0054 \end{bmatrix}^\top \quad \text{LC1} \\ \mathbf{y}_0 &= \begin{bmatrix} 5.9650 & -6.6938 & -4.5901 & 0.2163 & 8.6960 & -6.1122 \end{bmatrix}^\top \quad \text{LC2} \ . \end{aligned} \tag{A3}$$

References

1. Ibrahim, R.A. Friction-Induced Vibration, Chatter, Squeal, and Chaos—Part I: Mechanics of Contact and Friction. *Appl. Mech. Rev.* **1994**, *47*, 209. [CrossRef]
2. Kinkaid, N.M.; O'Reilly, O.M.; Papadopoulos, P. Automotive disc brake squeal. *J. Sound Vib.* **2003**, *267*, 105–166. [CrossRef]
3. von Wagner, U.; Hochlenert, D.; Hagedorn, P. Minimal models for disk brake squeal. *J. Sound Vib.* **2007**, *302*, 527–539. [CrossRef]
4. Hoffmann, N.; Gaul, L. Friction Induced Vibrations of Brakes: Research Fields and Activities. In *Friction Induced Vibrations of Brakes: Research Fields and Activities*; SAE Technical Paper Series; SAE International: Warrendale, PA, USA, 2008. [CrossRef]
5. Awrejcewicz, J.; Grzelcyk, D. Modeling and Analysis of Thermal Processes in Mechanical Friction Clutch—Numerical and Experimental Investigations. *Int. J. Struct. Stab. Dyn.* **2013**, *13*, 1340004. [CrossRef]
6. Ritto, T.G.; Escalante, M.R.; Sampaio, R.; Rosales, M.B. Drill-string horizontal dynamics with uncertainty on the frictional force. *J. Sound Vib.* **2013**, *332*, 145–153. [CrossRef]
7. Weiss, C.; Gdaniec, P.; Hoffmann, N.P.; Hothan, A.; Huber, G.; Morlock, M.M. Squeak in hip endoprosthesis systems: An experimental study and a numerical technique to analyze design variants. *Med. Eng. Phys.* **2010**, *32*, 604–609. [CrossRef]
8. Hoffmann, N.; Fischer, M.; Allgaier, R.; Gaul, L. A minimal model for studying properties of the mode-coupling type instability in friction induced oscillations. *Mech. Res. Commun.* **2002**, *29*, 197–205. [CrossRef]
9. Hoffmann, N. Transient Growth and Stick-Slip in Sliding Friction. *J. Appl. Mech.* **2006**, *73*, 642. [CrossRef]
10. Sinou, J.J.; Jézéquel, L. Mode coupling instability in friction-induced vibrations and its dependency on system parameters including damping. *Eur. J. Mech. A/Solids* **2007**, *26*, 106–122. [CrossRef]
11. Kruse, S.; Tiedemann, M.; Zeumer, B.; Reuss, P.; Hetzler, H.; Hoffmann, N. The influence of joints on friction induced vibration in brake squeal. *J. Sound Vib.* **2015**, *340*, 239–252. [CrossRef]
12. Papangelo, A.; Ciavarella, M.; Hoffmann, N. Subcritical bifurcation in a self-excited single-degree-of-freedom system with velocity weakening–strengthening friction law: Analytical results and comparison with experiments. *Nonlinear Dyn.* **2017**, *90*, 2037–2046. [CrossRef]
13. Jahn, M.; Stender, M.; Tatzko, S.; Hoffmann, N.; Grolet, A.; Wallaschek, J. The extended periodic motion concept for fast limit cycle detection of self-excited systems. *Comput. Struct.* **2019**, 106–139. [CrossRef]
14. Stender, M.; Jahn, M.; Hoffmann, N.; Wallaschek, J. Hyperchaos co-existing with periodic orbits in a frictional oscillator. *J. Sound Vib.* **2020**, *472*, 115–203. [CrossRef]
15. Habib, G.; Cirillo, G.I.; Kerschen, G. Uncovering detached resonance curves in single-degree-of-freedom systems. *Procedia Eng.* **2017**, *199*, 649–656. [CrossRef]
16. Gräbner, N.; Tiedemann, M.; von Wagner, U.; Hoffmann, N. *Nonlinearities in Friction Brake NVH-Experimental and Numerical Studies*; SAE: Warrendale, PA, USA, 2014.
17. Papangelo, A.; Hoffmann, N.; Grolet, A.; Stender, M.; Ciavarella, M. Multiple spatially localized dynamical states in friction-excited oscillator chains. *J. Sound Vib.* **2018**, *417*, 56–64. [CrossRef]
18. Awrejcewicz, J. Chaos in simple mechanical systems with friction. *J. Sound Vib.* **1986**, *109*, 178–180. [CrossRef]
19. Coudeyras, N.; Sinou, J.J.; Nacivet, S. A new treatment for predicting the self-excited vibrations of nonlinear systems with frictional interfaces: The Constrained Harmonic Balance Method, with application to disc brake squeal. *J. Sound Vib.* **2009**, *319*, 1175–1199. [CrossRef]

20. Gdaniec, P.; Weiß, C.; Hoffmann, N. On chaotic friction induced vibration due to rate dependent friction. *Mech. Res. Commun.* **2010**, *37*, 92–95. [CrossRef]
21. de Witte, V.; Govaerts, W.; Kuznetsov, Y.; Meijer, H. Analysis of bifurcations of limit cycles with Lyapunov exponents and numerical normal forms. *Phys. D Nonlinear Phenom.* **2014**, *269*, 126–141. [CrossRef]
22. Hetzler, H.; Schwarzer, D.; Seemann, W. Analytical investigation of steady-state stability and Hopf-bifurcations occurring in sliding friction oscillators with application to low-frequency disc brake noise. *Commun. Nonlinear Sci. Numer. Simul.* **2007**, *12*, 83. [CrossRef]
23. Hetzler, H. Bifurcations in autonomous mechanical systems under the influence of joint damping. *J. Sound Vib.* **2014**, *333*, 5953–5969. [CrossRef]
24. Grolet, A.; Thouverez, F. Computing multiple periodic solutions of nonlinear vibration problems using the harmonic balance method and Groebner bases. *Mech. Syst. Signal Process.* **2015**, *52–53*, 529–547. [CrossRef]
25. Hoffmann, N. Linear stability of steady sliding in point contacts with velocity dependent and LuGre type friction. *J. Sound Vib.* **2007**, *301*, 1023–1034. [CrossRef]
26. Hetzler, H.; Schwarzer, D.; Seemann, W. Steady-state stability and bifurcations of friction oscillators due to velocity-dependent friction characteristics. *Proc. Inst. Mech. Eng. Part J J. Eng. Tribol.* **2007**, *221*, 401–412. [CrossRef]
27. Nakano, K.; Maegawa, S. Safety-design criteria of sliding systems for preventing friction-induced vibration. *J. Sound Vib.* **2009**, *324*, 539–555. [CrossRef]
28. Stender, M.; Tiedemann, M.; Hoffmann, L.; Hoffmann, N. Determining growth rates of instabilities from time-series vibration data: Methods and applications for brake squeal. *Mech. Syst. Signal Process.* **2019**, *129*, 250–264. [CrossRef]
29. Stender, M.; Di Bartolomeo, M.; Massi, F.; Hoffmann, N. Revealing transitions in friction-excited vibrations by nonlinear time-series analysis. *Nonlinear Dyn.* **2019**, *47*, 209. [CrossRef]
30. Milnor, J. On the concept of attractor. *Commun. Math. Phys.* **1985**, *99*, 177–195. [CrossRef]
31. Strogatz, S.H. *Nonlinear Dynamics and Chaos: With Applications to Physics, Biology, Chemistry, and Engineering*, 2nd ed.; Studies in Nonlinearity; Perseus Books: Cambridge, MA, USA, 2001.
32. Menck, P.J.; Heitzig, J.; Marwan, N.; Kurths, J. How basin stability complements the linear-stability paradigm. *Nat. Phys.* **2013**, *9*, 89–92. [CrossRef]
33. Schultz, P.; Menck, P.J.; Heitzig, J.; Kurths, J. Potentials and limits to basin stability estimation. *New J. Phys.* **2017**, *19*, 023005. [CrossRef]
34. Mitra, C.; Choudhary, A.; Sinha, S.; Kurths, J.; Donner, R.V. Multiple-node basin stability in complex dynamical networks. *Phys. Rev. E* **2017**, *95*, 032317. [CrossRef] [PubMed]
35. Rakshit, S.; Bera, B.K.; Majhi, S.; Hens, C.; Ghosh, D. Basin stability measure of different steady states in coupled oscillators. *Sci. Rep.* **2017**, *7*, 45909. [CrossRef] [PubMed]
36. Stender, M.; Hoffmann, N. bSTAB (V1). 2020. Available online: https://www.preprints.org/manuscript/202011.0234/v1 (accessed on 6 November 2020). [CrossRef]

Publisher's Note: MDPI stays neutral with regard to jurisdictional claims in published maps and institutional affiliations.

lubricants

MDPI

Article

Numerical and Experimental Analysis of Nonlinear Vibrational Response due to Pressure-Dependent Interface Stiffness

Dorra Nouira [1,2], Davide Tonazzi [2], Anissa Meziane [1], Laurent Baillet [3] and Francesco Massi [2,*]

1 Institute of Mechanical Engineering, I2M-UMR 5295 CNRS-University of Bordeaux, 351 cours de liberation-Bat A4-RDC, 33405 Talence, France; dorra.nouira@u-bordeaux.fr (D.N.); anissa.meziane@u-bordeaux.fr (A.M.)

2 DIMA-Dipartimento di Ingegneria Meccanica e Aerospaziale, Sapienza University of Rome, Via Eudossiana 18, 00184 Rome, Italy; davide.tonazzi@uniroma1.it

3 Institute of Earth Sciences (ISTerre) BP 53, CEDEX 9, 38041 Grenoble, France; laurent.baillet@univ-grenoble-alpes.fr

* Correspondence: francesco.massi@uniroma1.it; Tel.: +39-06-44-58-58-90

Received: 9 June 2020; Accepted: 7 July 2020; Published: 10 July 2020

Abstract: Modelling interface interaction with wave propagation in a medium is a fundamental requirement for several types of application, such as structural diagnostic and quality control. In order to study the influence of a pressure-dependent interface stiffness on the nonlinear response of contact interfaces, two nonlinear contact laws are investigated. The study consists of a complementary numerical and experimental analysis of nonlinear vibrational responses due to the contact interface. The laws investigated here are based on an interface stiffness model, where the stiffness property is described as a nonlinear function of the nominal contact pressure. The results obtained by the proposed laws are compared with experimental results. The nonlinearity introduced by the interface is highlighted by analysing the second harmonic contribution and the vibrational time response. The analysis emphasizes the dependence of the system response, i.e., fundamental and second harmonic amplitudes and frequencies, on the contact parameters and in particular on contact stiffness. The study shows that the stiffness–pressure trend at lower pressures has a major effect on the nonlinear response of systems with contact interfaces.

Keywords: nonlinear dynamic response; second harmonics; experiments; numerical modelling; interface stiffness

1. Introduction

Accurate contact interface modelling requires a knowledge of interfacial parameters, including interface contact stiffness. For many applications, characterizing and understanding the contribution of the interface to the dynamic response of the system is critical. These include robotic applications [1], grippers [2], micro-bearings [3], adhesive surfaces [4] and, wherever dry contact occurs between solids [5], with specific attention to lightly loaded joints. In the case of structural diagnostic, health monitoring and quality control of components and joints, these are based on the measurement and interpretation of wave interaction with joint interfaces or component defects [6].

Although contacts are common in practical engineering applications, there are certain aspects, such as sensitivity to interfacial parameters, which are not fully understood and modelled. Such sensitivity causes uncertainty in system performances and reliability predictions.

One of the main methods used to model a contact interface is to use a spring and a viscous damper in parallel. Contact stiffness can be obtained from analytical contact models, for instance, the Herzian contact model for spherical contacts [7]. In the case of rough surfaces in contact, the Greenwood and Williamson [8] statistical model and its successive reformulations [6,9–11] have been used to obtain overall mean stiffness. Experimental values have been extracted using indirect methods [12] or system identification methods [13].

Recently, Jin et al. [14] used a quasi-static model developed within the GW framework, in which all the microscopic geometric features of contact interfaces are extracted directly from high-resolution scanning electron microscopy (SEM) images of real fatigue cracks.

However, the development of increasingly sophisticated numerical models with contact interfaces means that more reliable and fine contact parameters need to be defined. Contact stiffness has been proved to be sensitive to contact conditions such as contact pressure [15–17], third body rheology [18] and the true area of contact [19].

In more detail, the force is supported by surface asperities. As the force increases, more asperities come into contact, while each asperity undergoes flattening deformation. In [20], three contact states can be identified: total sliding, partial slip and contact loss. In the case of partial slip, roughness has been described by Aleshin [20] using the Method of Memory Diagrams (MMD), a model developed to describe partial slip for rough surfaces in contact. The MMD model was then extended to take into account the other two regimes of total sliding and contact loss [21,22]. The contact interface has a further nonlinear behaviour due to asymmetry between traction [23,24] and compression configurations. During compression, the change in the contact interface configuration, as a function of contact pressure, also results in nonlinearity in the interface response.

When these nonlinearities are activated by the interaction between propagating waves and the contact interface, higher-order harmonics are then generated [25]. While these effects have been well studied in the ultrasonic field [26], they also represent a new area of investigation from a vibrational point of view [9]. In particular, the generation and features of second harmonics [27] deserve to be further analysed and exploited.

In this context, the aim of this study is to present a numerical and experimental analysis to provide a basic insight into the nonlinear vibrational response of a contact interface, as a basis for evaluating nonlinear contact through stress-dependent stiffness in compression.

To this end, a numerical model with contact interfaces was developed, considering different nonlinear contact models, with different stress-dependent stiffnesses. A specific contact law is proposed, including a specific evolution of the stiffness for low pressures, and compared to a classical power law, fitting experimental values.

An experimental campaign was then conducted on a specific test bench in order to investigate the nonlinear response of the system, tested under a contact pressure of up to 1 MPa. By comparing experimental and numerical nonlinear responses, the sensitivity of the system response to the contact interface stiffness trend at low pressures was highlighted, where experimental data are missing in the literature.

2. Materials and Methods

2.1. Description of the Approach

The study combines numerical and experimental analyses to provide basic insights into the nonlinear dynamic response of a system with contact interfaces, under propagating waves due to an impulsive-type force.

The nonlinear response of an experimental test bench (Figure 1) to an impulsive force is then modelled in a one-dimensional framework, including two rough contact interfaces with nonlinear contact stiffness.

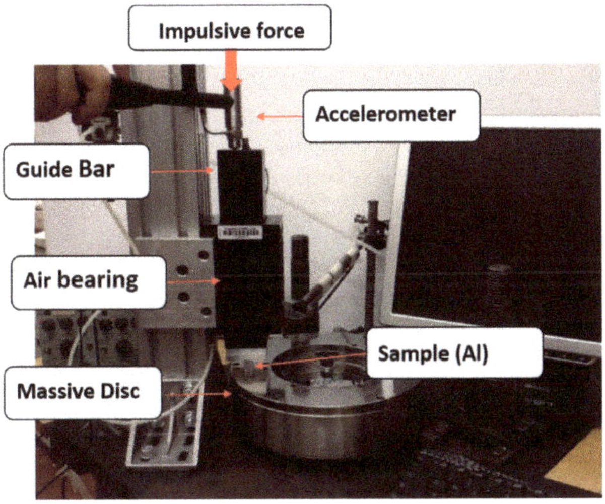

Figure 1. Experimental set-up.

The tested contact stiffness laws are first identified over a large contact pressure range, using both the experimental tests and results from the literature [15,18,28]. However, there is a lack, both experimentally and from the literature, of assessments of the stiffness trend within the lower pressure range (less than 0.14 MPa). By comparing the nonlinear time responses of the experimental system with the numerical results obtained by different stiffness–pressure curves, this gap is discussed.

The analysis was conducted in terms of the evolution of the amplitude of the fundamental and second harmonics and their frequencies, as a function of the force amplitude. In fact, due to the nonlinearity of the contact in compression, higher harmonics [26] appear in the spectrum of the vibrational response.

The numerical results derived from a specific nonlinear law, reconstructed from the available experimental data, are compared to other contact laws in the literature, in particular with respect to constant linear contact stiffness [29] and the power–law relation between stiffness and pressure [15].

In this article, the system remains in compression for all the configurations throughout the simulations. The "clapping" effect, which leads to a further strong nonlinearity [30] and denotes intermittent loss of contact at the interface, is not studied in this work.

2.2. Experimental Set-up

The set-up used for all the experimental measurements is presented in Figure 1. The system was designed to estimate the contact stiffness between two rough interfaces of different sample materials, within a range of average contact pressure up to 1 MPa, in both sticking and sliding conditions [18]. The system consists of a sample in contact with a massive steel disc and loaded by the dead weights on a guide bar, as shown in Figure 1.

The guide is maintained by an air bearing, to enable it to oscillate and not introduce further stiffness and friction in the vertical direction.

The tested sample consists of aluminium (Al). The material properties and surface roughness are presented in Table 1.

Table 1. The tested aluminium sample.

Individual Sample Parameters				Interface Parameter
Length (m)	Contacting surface (m^2)	Young modulus (GPa)	Density (kgm^{-3})	Roughness (μm) Ra
0.015	1.15×10^{-4}	71	2710	1

An impulsive-type force is applied by an instrumented impact hammer (Brüel and Kjær type 8202), while the dynamic response of the system is recorded by an accelerometer placed on top of the guide bar.

All tests presented in the paper are performed on the system, with the overall weight of the guide bar on top of the sample, generating a static equilibrium pressure $\sigma_0 = 0.14$ MPa.

Measurements of the force and acceleration of the system are recorded using the acquisition system (SIRIUS–DEWESOFT), based on DualCoreADC® technology with dual 24-bit delta-sigma analogue to digital converter (ADC). An anti-aliasing filter on each analogue channel achieves a 160 dB dynamic range in time and frequency domains with 200 kHz sampling rate per channel. Data are post-processed by Matlab.

We assume that there is no interface separation between the two contact surfaces during the tests. This assumption is supported by the numerical simulations for the levels of impulsive excitation used.

2.3. D Numerical Model

A one-dimensional numerical model of the experimental set-up was implemented (Figure 2) consisting of the guide bar Ω_1 of the experimental set-up and the tested aluminium sample Ω_2, modelled by unidimensional deformable bodies.

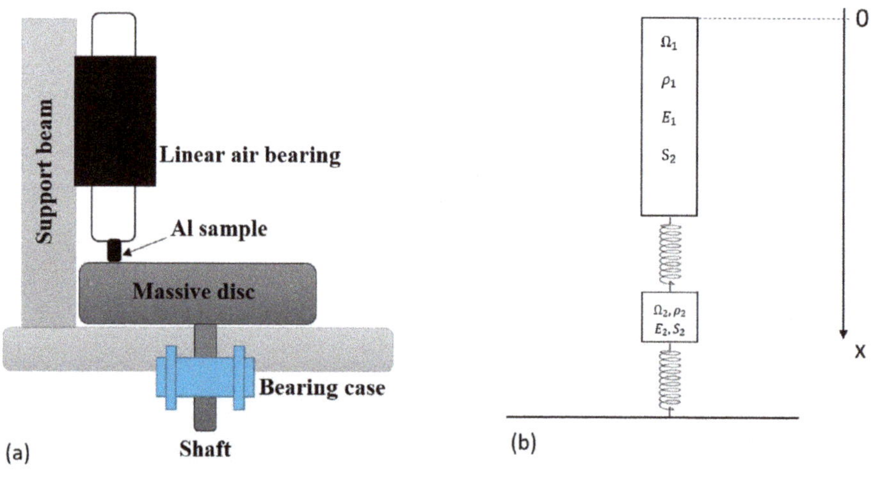

Figure 2. Diagram (**a**) and numerical model (**b**) of the set-up for contact stiffness experiments.

Two contacting interfaces are considered: the first between the guide bar Ω_1 and the tested sample Ω_2, and the second between the tested sample and the frame, considered as infinitely rigid in the model (tribometer disc). The model parameters are provided in Table 2. Note that, in reality, the cross-sections of the guide bar Ω_1 and aluminium sample Ω_2 are different. Numerically, an equivalent 1D-model with the same cross-section S_2 is considered, corresponding to the contact surface S_2.

Table 2. Geometry and material parameters of Ω_1 and Ω_2 in the numerical model.

	Length L (m)	Mass m (Kg)	Young's Modulus E (GPa)	Density (kgm^{-3})	Contacting Surface S_2 (m^2)
Guide bar (Ω_1)	0.229	1.61	71	2710	1.15×10^{-4}
Sample (Ω_2)	0.015	0.00467	71	2710	1.15×10^{-4}

For both interfaces, a nonlinear contact law is introduced. This law includes a nonlinear stiffness function of the contact pressure.

Once a force is applied to the top of the system, longitudinal waves propagate along the vertical x-axis and excite the longitudinal modes of the system, including the mass-spring mode, where the mass is the guide bar and the spring is due to the stiffness in series of the sample and the two contact interfaces.

The numerical modelling assumes the following hypotheses:

- The dynamic of interest is in the vertical direction, i.e., the numerical model can be reduced to one dimension (1D) (Figure 2b);
- The massive disc in the set-up can be considered as rigid within the frequency range of interest.
- Throughout the analysis, the interface remains in contact. Thus, no dissipation occurs at the interface and damping at the contact is disregarded;
- The impact of the hammer can be modelled by introducing the respective force F into the boundary conditions, measured at the tip of the instrumented hammer:

$$\sigma_{/x=0} = \frac{F}{S_2}$$

where $\sigma_{/x=0}$ is the stress at the upper side of the guide bar.

a. Governing equations

In the absence of elastic wave, the system is assumed to be free at the upper part and subject only to its own weight. In a first static step, the equilibrium position of the system can be calculated.

For the longitudinal waves propagating along the x-direction (longitudinal), in both the guide bar and the aluminum sample, the equation of motion is the following

$$\frac{\partial^2 u}{\partial t^2} - c^2\frac{\partial^2 \mathrm{u}}{\partial x^2} + h\frac{\partial \mathrm{u}}{\partial t} = g \tag{1}$$

where $u\ (x,t)$ is the displacement in the x-direction at time t, from the non-deformed configuration, c is the wave velocity, g is the gravity acceleration, h is the viscosity factor and t denotes the time.

The viscosity factor h is constant in the model and was determined from the damping factor $\xi = \frac{\mathrm{h}}{2m\omega_0}$, where ω_0 is the angular frequency of interest. The damping factor value $\xi = 0.035$ was identified experimentally by logarithm decrement at the frequency of interest, which is the frequency of the corresponding mass-spring mode. In fact, the main mode of vibration, the focus of the analysis, is the response of the guide mass to the stiffness obtained by the bulk stiffness of the sample in series with the stiffness of the two contact interfaces (Figure 2).

The model is discretized in both time and space using the Euler scheme [31].

The unidimensional wave Equation (1) has been discretized using a second order centered finite difference (FD) scheme, explicit in time, as follows

$$\begin{aligned}\frac{u_k^{m+1}-2u_k^m+u_k^{m-1}}{\delta t^2} \quad & -c^2\frac{u_{k+1}^m-2u_k^m+u_{k-1}^m}{\delta x^2} + h\frac{u_k^{m+1}-u_k^m}{\delta t} \\ & = g\,,\ \forall k\epsilon[2, n_1-1] \cup [n_1+2, n_1+n_2-1]\end{aligned} \tag{2}$$

where the subscript k indicates the node number and m indicates the time step number, with δx and δt corresponding, respectively, to the space and time discretisation steps. n_1 and n_2 are, respectively, the total number of nodes contained in solids Ω_1 and Ω_2. This second-order general scheme is conditionally stable under the classic Courant–Friedrichs–Lewy (CFL) condition:

$$\frac{c\delta t}{\delta x} < 1 \tag{3}$$

Equation (2) can be re-written to obtain the displacement at the next time step $m + 1$ for each internal node, as follows

$$u_k^{m+1} = u_k^m\left(\frac{2(1-\alpha_1)+\alpha_2}{1+\alpha_2}\right) + \frac{\alpha_1}{1+\alpha_2}\left(u_{k+1}^m + u_{k-1}^m\right) + \frac{1}{1+\alpha_2}\left(gdt^2 - u_k^{m-1}\right) \tag{4}$$

$$\forall k\epsilon[2, n_1 - 1] \cup [n_1 + 2, n_1 + n_2 - 1]$$

where $\alpha_1 = \frac{c^2\delta t^2}{\delta x^2}$ and $\alpha_2 = h.\,dt$

At contact interfaces (at x = x_{C1} between the two solids Ω_1 and Ω_2 and x = x_{C2} between Ω_2 and the frame), the contact stresses σ(x = x_{C1},t) and σ(x = x_{C2},t) has the following expression in compression

$$\sigma(x_{Ci}, t) = K(\sigma(x_{Ci}, t))\Delta u_{Ci} \tag{5}$$

where K denotes the nonlinear stiffness of the contact interface depending on the contact stress and Δu_{Ci} the gap distance at interfaces. i indicates either the contact between the two solids Ω_1 and Ω_2 (i = 1) or the contact between Ω_2 and the frame (i = 2).

In order to evaluate properly the solution at both contact interfaces ($\sigma(x_{Ci}, t)$, $K(\sigma(x_{Ci}, t))$ and Δu_{Ci}), it is necessary to solve Equation (5), and thus apply an implicite scheme (finite difference), using the relation between the contact pressure and the gap distance extracted from the implemented nonlinear relation between the stiffness and the pressure, presented in what follows.

b. Contact law

Contact between two interfaces is generally modelled by a relationship between the interfacial gap and the contact pressure. The contact laws can be extracted analytically from various statistical models of rough surfaces. Drinkwater [11] and Baltazar et al. [32] attempted to link the roughness topography to the contact stiffness, transmission/reflexion coefficients, measured ultrasonically, and frequency.

However, it is generally difficult to take into account all the detailed information about third body [33] features, i.e., local deformations and interactions within a real interface and roughness [34,35].

As an alternative approach, the desired function can be modelled as a nonlinear contact stiffness, with respect to the applied contact pressure. When two rough surfaces are pressed together, the stiffness K_c per unit of area of the interface is given by the rate of change in nominal pressure σ with the average interfacial gap. Contact stiffness K_c can be determined in general from [9]

$$K_c = -\frac{\partial\sigma}{\partial u} \tag{6}$$

where σ is the nominal contact pressure and u is the average interfacial gap.

In the literature, for higher contact pressures (greater than 0.14 MPa, which is the lowest value of pressure used for the experimental estimation of contact stiffness, see Table 2), contact stiffness is often assumed to follow a power law [15]. This latter model gives the relationshi between the contact stiffness and the applied pressure as follows

$$K_c = -\frac{\partial\sigma}{\partial u} = -C\sigma^m \tag{7}$$

where C and m are positive constants.

In this study, the contact stiffness of the tested samples was previously determined by experimental measurements, as reported in [18]. From preliminary dynamic tests at different contact pressures, the contact stiffness between 0.14 and 1 MPa could be estimated. Table 3 shows the results for contact stiffness as a function of the average contact pressure for the aluminum–aluminum interface, with a surface roughness of Ra = 1 μm.

Table 3. Normal contact stiffness as a function of the average contact pressure in sticking condition [15].

	Aluminium Ra = 1 μm			
Contact pressure [MPa]	P = 0.14	P = 0.35	P = 0.57	P = 1
Kc [Pa/m]	1.15×10^{12}	1.75×10^{12}	1.63×10^{12}	2.46×10^{12}

The data highlight how the contact stiffness increases with the rise in the average contact pressure. The contact stiffness values range from 1.15 × 1012 to 2.46 × 1012 Pa/m when the contact pressure increases from 0.14 to 1 MPa.

Even if a contact interface is expected to show hysteresis, with a slightly different stiffness during loading and unloading phases, the authors considered this effect negligible, or at least not measurable. The complexity of an interface, including third body particles, chemical bounds and oxides, leads to the need for an overall approximation, which is here represented by the stiffness parameter. Of course, such approximation can have different implications, as a function of the wished phenomena to be investigated.

These experimental results are useful for defining the numerical contact stiffness within the tested range of contact pressures. Nevertheless, for implementing the contact law in the numerical simulations, it is necessary to define the stiffness for the whole pressure range [0; 1 MPa], in which the contact pressure will vary.

A first approximation within this pressure range was obtained by approximating the experimental measurements by a power law function, as in Equation (3). The least squares method allows a good agreement with the experimental data (Table 3), which are available from pressure 0.14 MPa. This agreement is obtained for $C = 1.81 \times 10^{10}$ Pa/m and m = 0.35. The power law ("PL" in Figure 3a) is thus defined to approximate the experimental points (Table 3).

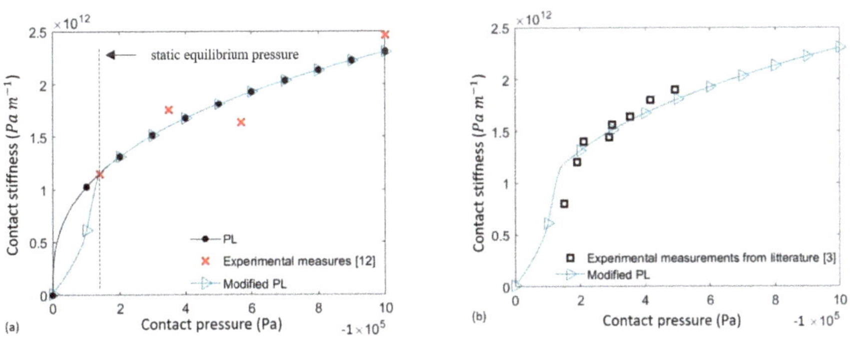

Figure 3. Normal contact stiffness as a function of contact pressure in compression conditions. (**a**) Experimental results (red cross); Modified PL (blue triangle); PL (black circle). (**b**) Modified PL (blue triangle) and experimental data from the literature [6] (black square)

While this power law has been built with consideration to the measurements and the literature dealing with higher pressures, the approximation of the trend at lower pressures is completely arbitrary. In order to model the contact stiffness trend for lower contact pressures, other experimental observations in the literature [6] have been exploited. These experimental results show the existence of an inflexion point for low-contact pressures (Figure 3b). From experimental results (Table 3) and the literature [6] it is then possible to propose a different overall contact stiffness trend, as a function of the contact pressure, including the inflection point between 0.14 and 0 MPa ("Modified PL" in Figure 3.

Finally, the stiffness laws, which will be investigated in the following, are:

- The "PL" law, corresponding to the best approximation of experimental data (Table 3) by a power law.
- The "Modified PL", defined piecewise. It is equal to "PL" for pressures greater than 0.2 MPa, and for lower pressures it accounts for the inflexion point reported in [6].

The relationship between the contact pressure and the interfacial gap can then be extracted from the implemented nonlinear relation between stiffness and pressure. For example, the results derived from the "Modified PL" are shown in Figure 4, highlighting the nonlinear response of the interface.

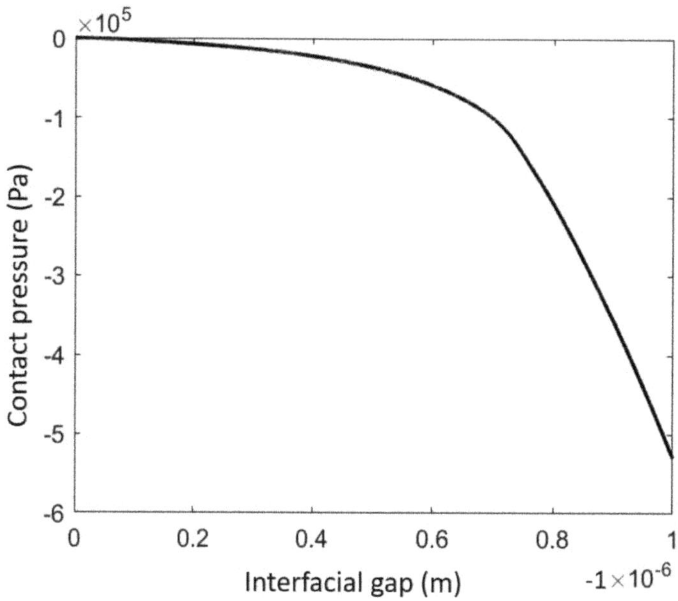

Figure 4. The corresponding contact pressure-interfacial gap trend for the 'Modified PL'.

In the following, the experimental nonlinear response of the system to an impulsive excitation force will be compared with the nonlinear response obtained by the simulations with the different laws analysed. The comparison will allow for a discussion of the different laws considered to simulate the effective interface stiffness nonlinearity.

3. Experimental and Numerical Comparison

3.1. Dynamic Response of the Contact System

The aim of this section is to compare the numerical and experimental dynamic responses of the system (Figure 2), in time and frequency. For the sake of conciseness, the comparison of the time signals with the experiments is first reported only for the "Modified PL"; in the following section, the general results obtained by both laws will be compared with the experiments.

Experimentally, an impulsive force was applied to the upper side of the guide by an instrumented hammer (PCB Piezotronics-086C03). The case presented hereafter corresponds to an impulsive force of 32 N. The applied force and the acceleration are measured and shown in Figure 5a,b, respectively. For the numerical model, the measured experimental force has been interpolated (see Figure 5a) and introduced as a boundary condition in the numerical simulation. The model presented in Section 2, with the "Modified PL", has been used to simulate the system response to the force.

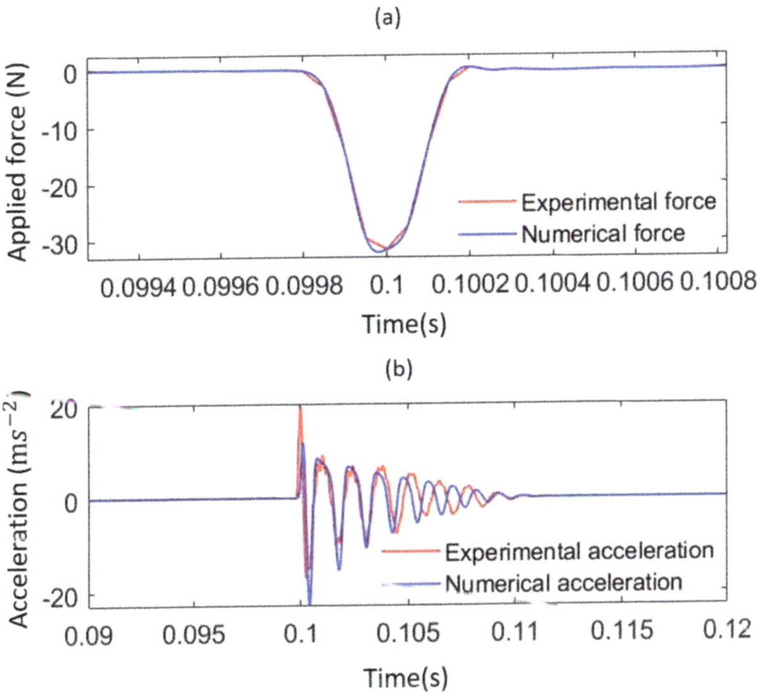

Figure 5. (**a**) Force signals over time, for a single force; (**b**) Acceleration signals over time obtained with 'Modified PL'. Test performed with average contact force 32 N.

Figure 5b shows the respective experimental and numerical accelerations, due to the dynamic system response.

Experimental and numerical responses show good agreement in amplitude and time evolution. Figure 6 shows the Frequency Response Functions (FRF) [26], which provide the response of a system to an external excitation in the frequency domain. They are calculated from both numerical and experimental signals, to characterize the dynamics of the system. The numerical curves shown in Figure 6 correspond to the one obtained with a constant interface stiffness of 8.5 × 1011 Pa/m (dashed line) and the one obtained with the "Modified PL" presented in Figure 3. The equivalent constant stiffness of 8.5 × 1011 Pa/m was calculated to obtain the same frequency for the first harmonics of the "Modified PL", to highlight its nonlinear contribution to the system response.

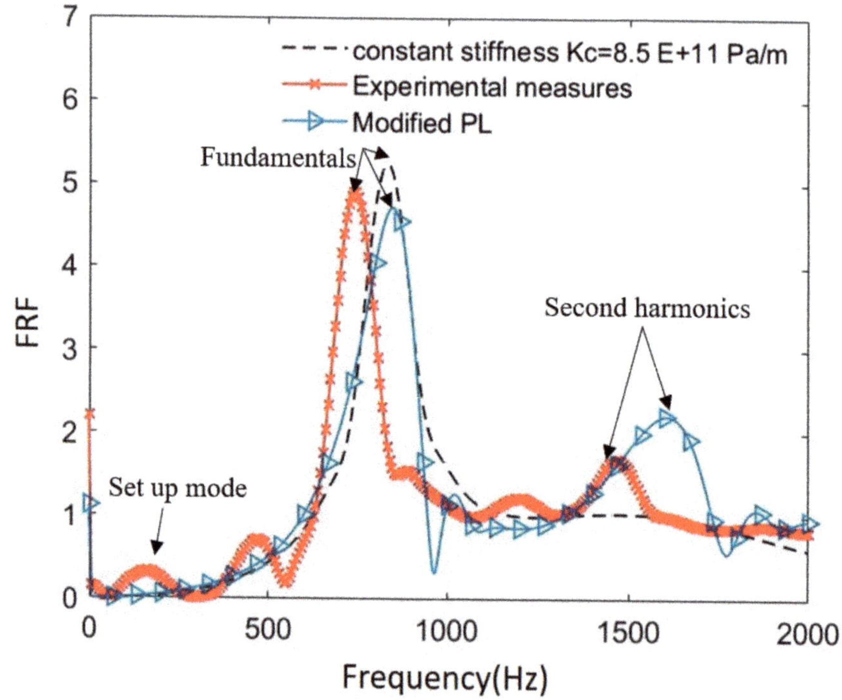

Figure 6. FRFs of the system (receptance, displacement/force). Numerical with nonlinear stiffness, numerical with constant stiffness with Kc = 8.5 × 10^{11} Pa/m and experimental. Test performed with average contact force 32 N.

Only the fundamental (f~800 Hz) and second harmonics (f~1600 Hz) of the mass-spring mode are investigated here. Their results are well decoupled from the rest of the system dynamics.

The numerical spectra, obtained with either constant or nonlinear stiffness, show a peak around frequency f_1 = 800 Hz, corresponding to the fundamental frequency of the system. This is the natural frequency of the mass-spring mode, where the mass is the guide, while the spring is the series of the two interfaces and the sample stiffness (Figure 2). The comparison shows good agreement between the numerical and experimental results in terms of frequency and width of the peaks, i.e., damping. The amplitude of the fundamental is well simulated as well, with a percentage error less than 10%.

Unlike the spectra obtained with the constant stiffness, those based on nonlinear stiffness show a peak around frequency $f_2 = 2f_1$ = 1600 Hz, which is also recovered experimentally. This peak represents the second harmonic and correlates with the experimental second harmonic.

The presence of the second harmonic in the spectra is due to the nonlinear nature of the contact stiffness. This is confirmed by the absence of this harmonic in the numerical results obtained with the constant contact stiffness (Figure 6 dashed line), and thus the occurrence of the second harmonic in experiments can be directly correlated with the nonlinearity of the interface stiffness.

3.2. Nonlinear Response of the Interface

In order to discuss the proposed trends of numerical contact stiffness and evaluate it by comparison with experiments, a spectrum analysis of the acceleration signals is reported in this section, as a function of the amplitude of the impulsive force. It is assumed that an increase in the force, and then in the system response, increases the nonlinear contribution of the interface to the system response.

A comparison of experimental and numerical FRFs, derived from the "Modified PL", is carried out for different impulsive force amplitudes, ranging from 9 to 32 N (Figure 7).

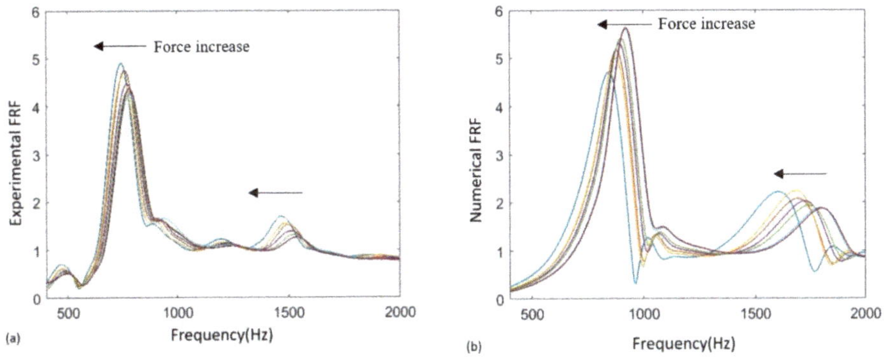

Figure 7. Frequency response functions (FRFs) of the system (receptance, displacement/force). (**a**) Experimental frequency response; (**b**) Numerical frequency response plotted with "Modified PL". Test performed with average contact force ranging from 9 to 32 N.

When increasing the force amplitude, the overall average stiffness at the interface decreases, leading to a decrease in mode frequency (Figure 7). The nonlinearity of the interface stiffness is observable by the appearance of the second harmonic (frequency from 1400 to 1800 Hz, depending on the amplitude of the impulsive force) in the system response.

In the following, the numerical results obtained with the different contact laws, as presented in Figure 3, are compared with the experimental results in terms of the magnitude of the fundamental and second harmonics, as well as in terms of the frequency of the fundamental, as a function of the applied force.

Figure 8 shows the frequency evolution of the fundamental harmonic as a function of the amplitude of the applied force. As observed in the experimental results, a decrease in frequency was obtained in the numerical simulations with the implemented "Modified PL". This decrease is due to the decrease in the effective average stiffness when the oscillation at the contact increases. Experimentally, this trend has already been observed in [18]. It is worth mentioning that the decrease in frequency when the force amplitude increases (Figure 8) is also recovered by the "PL", but with a lower slope. A slight decrease in frequency (2%) can be observed in Figure 8 for the "PL", which is lower than for the experimental one (5%). Conversely, the "Modified PL" introduces a greater decrease in terms of stiffness, particularly for low contact pressures, as shown in Figure 9.

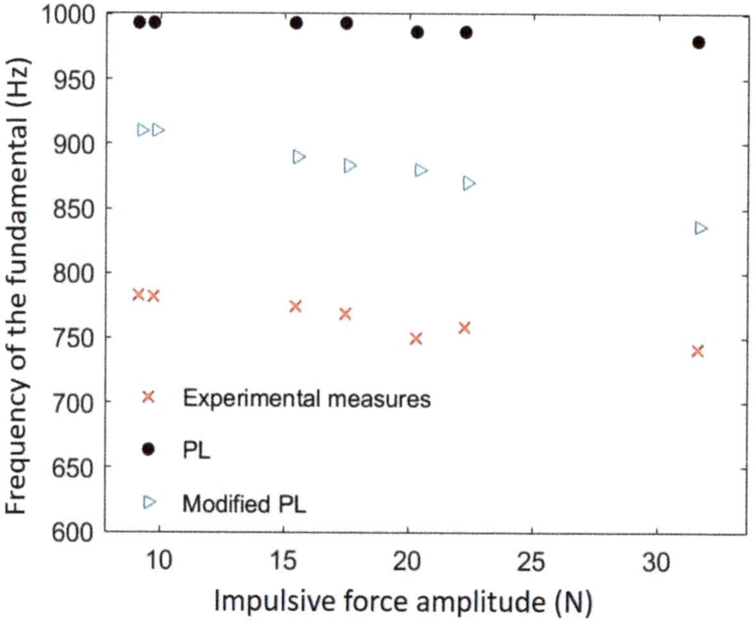

Figure 8. Frequency of the fundamental, as a function of the force amplitude [N]. Experimental measurements (red cross), PL (black circle) and Modified PL (blue triangle).

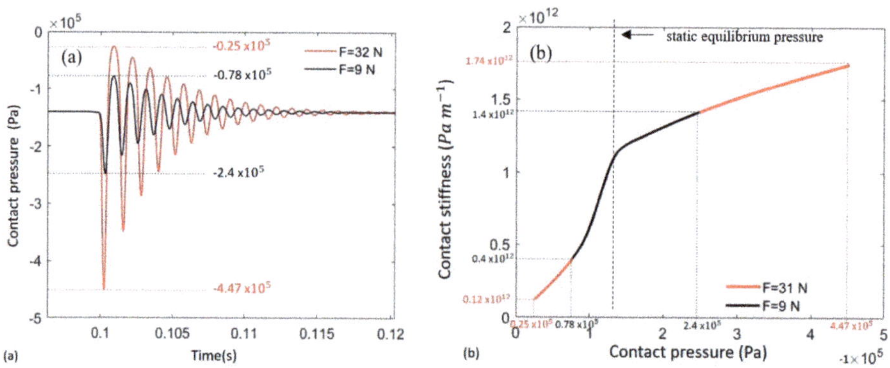

Figure 9. Numerical results for maximum force F = 32 N (red) and minimum force F = 9 N (black). (**a**) Numerical contact pressure as a function of time; (**b**) Numerical stiffness as a function of contact pressure. Test performed on aluminium with Modified PL.

It should also be noted that, while the trend of the frequency is correctly simulated by the proposed laws, an error in the absolute value of the frequency, around 13%, is observed. This is due to the non-infinite stiffness of the counterpart (tribometer disc), unlike the infinite stiffness in the simulation, which implies a lower experimental frequency. Thanks to the numerical results, the decrease in the frequency of the fundamental can be shown to be related to a decrease in the mean value of contact stiffness, with the increase in the applied force

$$K_{avr} = \frac{1}{T}\sum_{t=0}^{T}|K_c(t)| \tag{8}$$

where K_c is the stiffness and T is the time of simulation.

In order to highlight the difference in average stiffness according to the force amplitude, Figure 9a shows the system response to two different force amplitudes, using the "Modified PL". An applied force of 9 N generates a maximum contact pressure of −0.078 MPa, resulting in an average contact stiffness of 0.4 × 1012 Pa/m, while a force of 32 N generates a maximum contact pressure of −0.025 MPa and an average contact stiffness of 0.12 × 1012 Pa/m.

Figure 10 shows the evolution of average contact stiffness K_{avr} for the "Modified PL" and the "PL". It confirms the decrease in average stiffness when the applied force is increased, which explains the decrease in frequency (Figure 7).

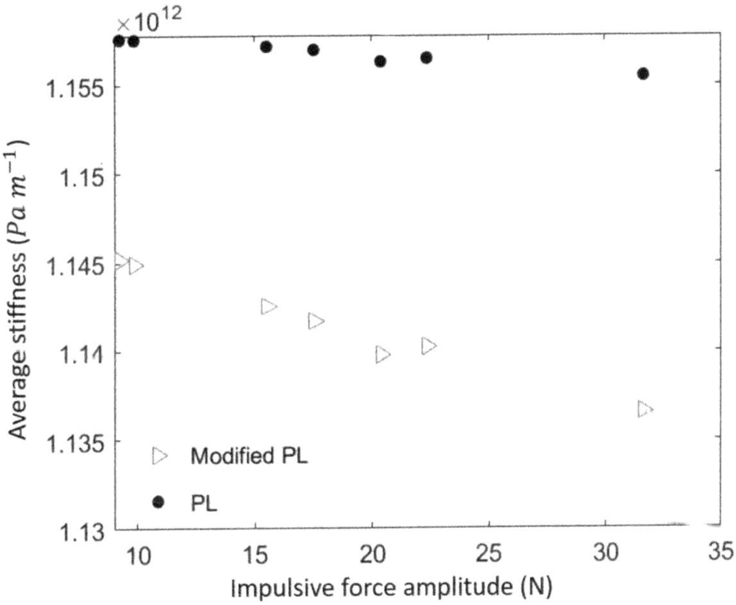

Figure 10. Average contact stiffness as a function of the applied forces. The test performed with average contact force ranging from 9 to 32 N, with Modified PL (blue triangle) and PL (black circle).

Figure 11 shows the evolution of the amplitude of the fundamental, as a function of the applied force, for the different contact laws.

Considering the mean value of the fundamental harmonic over the considered range of pressure, the "Modified PL" produces amplitudes closer to the experimental ones for this set of experimental measurements.

Nevertheless, as shown in Figure 11, for both of the implemented laws, the slight experimental increase in the amplitude of the fundamental (A1), with respect to the force amplitude, is not retrieved numerically. As mentioned above, this could be explained by the different boundary conditions between the numerical and experimental systems. In fact, the experimental set-up is not completely rigid, due to the deformability of the bench components (disc, shaft, bearings, etc.). Despite using a massive disc to isolate the dynamics of the investigated system (air guide and samples in contact) from the rest of the set-up as much as possible, a slight error is introduced by the residual flexibility of the system.

Finally, Figure 12 shows the ratio (A_2/A_1) of the amplitudes of the second harmonic (A_2) to the fundamental (A_1), obtained both experimentally and numerically, for both contact laws. It can be noted that the amplitude of the second harmonic is normalized by the amplitude of the fundamental, which is dependent on the energy introduced by the external force at this frequency, in order to highlight the nonlinear contribution originated by the contact interface.

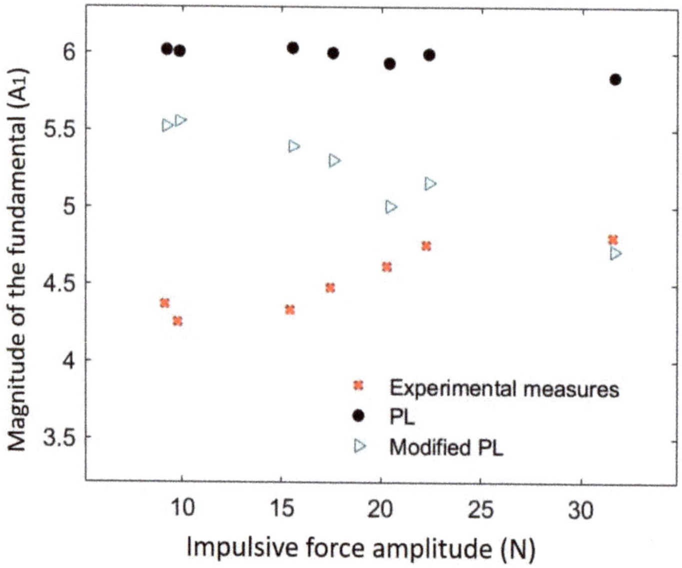

Figure 11. Magnitude of the FRF of the fundamental A1, as a function of the force amplitude. Comparison of the contact laws. Experimental measurements (red cross), Modified PL (blue triangle), and PL (black circle).

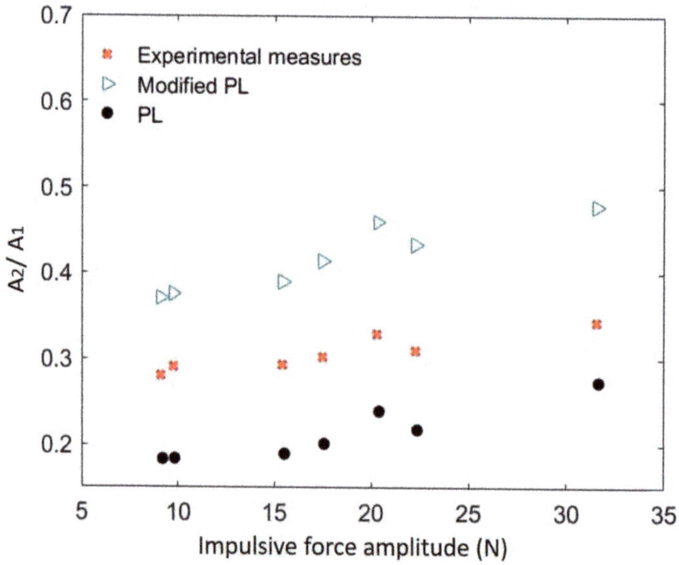

Figure 12. Ratio of magnitudes of the FRF of the second harmonic (A_2) to fundamental (A_1), as a function of the force amplitude. Experimental measurements (red cross), Modified PL (blue triangle) and PL (black circle).

The trends of the A_2/A_1 ratio, calculated for all the tested contact laws, are similar to the experimental trend. In general, an increase in the amplitude of the applied force (x axis in Figure 12), and consequently a higher amplitude of the system vibrational response, generates a greater nonlinear contribution, both numerically and experimentally. In fact, a larger oscillation of the contact pressure (especially within the low-contact pressure range, absolute value [0; 0.14 MPa]) generates a more nonlinear response by the system, which leads to a higher distortion of the signals and then a higher second harmonic contribution. The higher amplitude observed for the second harmonic of the "Modified PL" is due to the higher nonlinearity of the stiffness around the equilibrium position, with respect to the "PL".

The overall comparison, based on the analysis reported above, highlights the fact that the contact interface response depends heavily on the stiffness trend at lower pressures (less than 1 MPa). This stiffness trend at lower pressures introduced by the "Modified PL" increases the nonlinearity of the response (second harmonic amplitude) and decreases the average stiffness, i.e., the frequency of the main harmonics.

These results demonstrate that the stiffness trend at lower pressures plays a vital role and should be clearly identified, as it has a huge effect on the nonlinear response of mechanical systems with contact interfaces.

4. Conclusions

The nonlinear normal stiffness of contact interfaces, due to surface roughness, is a topic of major interest in several areas of application. A consequence of such nonlinearity is the appearance of second harmonic terms, either in acoustic wave propagation through the interface or in the dynamic vibrational response of systems with contact interfaces.

While contact stiffness nonlinearity at higher pressures has been widely discussed in the literature, and generally approximated by a power law, the contact stiffness trend at lower pressures has not been clearly identified. In this paper, a classical power law, fitted from experimental data at high contact pressures, is compared with a modified power law implementing an inflection point at lower pressures, where experimental data are not available. The stiffness-pressure trend within the higher contact pressure range was approximated from experimental measurements performed on a dedicated test bench. Within the lower contact pressure range, data from the literature were used to assume the different possible trends.

The nonlinear response of the system, obtained experimentally when exciting a dedicated system with an impulsive force, was analysed and compared with the nonlinear response of the numerical model that was developed, with the contact interface modelled by the different contact laws.

From the numerical simulations it was possible to identify the effect of the contact nonlinearity on the dynamic response of the system. The decrease in the average contact stiffness with the increase in the impulsive force explains the appearance of the second harmonics and the decrease in the fundamental frequency. In addition, the amplitude of the second harmonic was simulated and explained by the stiffness trend at the contact interface during the system oscillations.

In the nonlinear system response, the key role of the contact stiffness trend within the lower pressure range is highlighted, demonstrating the need to identify such parameters with dedicated experimental tests. Nonlinear contact laws and their effect on the dynamics of a system ought to be further investigated by implementing the contact laws considered here in finite element codes, in order to consider more realistic structures and interfaces.

Author Contributions: Methodology, D.N., A.M., L.B. and F.M.; visualization, D.N.; investigation, D.N. and D.T.; data curation, D.N. and D.T.; writing—original draft preparation, D.N.; writing—review and editing, A.M., L.B. and F.M.; validation, A.M., L.B. and F.M.; supervision, A.M. and F.M. All authors have read and agreed to the published version of the manuscript.

Funding: This research received no external funding.

Conflicts of Interest: The authors declare no conflict of interest.

References

1. Armstrong-Hélouvry, B. Control of Machines with Friction. *J. Tribol.* **1992**, *114*, 637. [CrossRef]
2. Koc, M.; Eray, T.; Sümer, B.; Cerci, N. An active force controlled laparoscopic grasper by using a smart material actuation. *Tribol. Int.* **2016**, *100*, 317–327. [CrossRef]
3. Bhushan, B. *Tribology and Mechanics of Magnetic Storage Devices*; Springer Science & Business Media: New York, NY, USA, 2012.
4. Türkmen, D.; Koc, I.M.; Sümer, B. Adhesive Pillar Based Air Levitation System for Contactless Manipulation of Fine Objects. *Procedia Eng.* **2015**, *120*, 1124–1127. [CrossRef]
5. Tonazzi, D.; Massi, F.; Baillet, L.; Brunetti, J.; Berthier, Y. Interaction between contact behaviour and vibrational response for dry contact system. *Mech. Syst. Signal Process.* **2018**, *110*, 110–121. [CrossRef]
6. Shi, X.; Polycarpou, A. Measurement and modeling of normal contact stiffness and contact damping at the meso scale. *J. Viration Acoust.* **2005**, *127*, 52–60. [CrossRef]
7. Johnson, K.L. *Contact Mechanics*; Cambridge University Press: Cambridge, UK, 1985.
8. Greenwood, J.A.; Williamson, J.P.B. Contact of nominally flat surfaces. *Proc. R. Soc. Lond.* **1966**, *295*, 300–319.
9. Richardson, J.M. Harmonic Generation at an Unbonded Interface: I.Planar Interface Between Semi-Infinite Elastic Media. *Int. J. Eng. Sci.* **1979**, *17*, 73–85. [CrossRef]
10. Rudenko, V. Nonlinear Acoustic Properties of a Rough Surface Contact and Acoustodiagnostics of a Roughness Height Distribution. *Acoust. Phys.* **1994**, *40*, 593–596.
11. Drinkwater, B.W.; Dwyer-Joyce, R.S.; Cawley, P. A Study of the Interaction Between Ultrasound and a Partially Contacting Solid-Solid Interface. *Proc. R. Soc. Lond.* **1996**, *452*, 2613–2628.
12. Polycarpou, A.; Soom, A. Boundary and Mixed Friction in the Presence of Dynamic Normal Loads. I. System Model. *J. Tribol.* **1995**, *117*, 261–266. [CrossRef]
13. Song, Y.; Hartwigsen, C.; McFarland, D.; Vakakis, A.; Bergman, L. Simulation of dynamics of beam structures with Bolted Joints using adjusted Iwan beam elements. *J. Sound Vib.* **2004**, *273*, 249–276. [CrossRef]
14. Jin, J.; Johnson, P.; Shokouhi, P. An integrated analytical and experimental study of contact acoustic nonlinearity at rough interfaces of fatigue cracks. *J. Mech. Phys. Solids* **2020**, *135*, 103769. [CrossRef]
15. Biwa, S.; Nakajima, S.; Ohno, N. On the acoustic nonlinearity of solid-solid contact with pressure-dependent interface stiffness. *J. Appl. Mech.* **2004**, *71*, 508–515. [CrossRef]
16. Aleshin, V.V.; Papangelo, A. Friction-induced energy losses in mechanical contacts subject to random vibrations. *Int. J. Solids Struct.* **2020**, *190*, 148–155. [CrossRef]
17. Pastewka, L.; Prodanov, N.; Lorenz, B.; Müser, M.H.; Robbins, M.O.; Persson, B.N. Finite-size scaling in the interfacial stiffness of rough elastic contacts. *Phys. Rev. E.* **2013**, *87*, 062809. [CrossRef] [PubMed]
18. Tonazzi, D.; Massi, F.; Salipante, M.; Berthier, Y. Estimation of the normal contact stiffness for frictional interface in sticking and sliding condition. *Lubricants* **2019**, *7*, 56. [CrossRef]
19. Kendall, K.; Tabor, D. An ultrasonic study of the area of contact between stationary and sliding surfaces. *Proc. R. Soc. Lond.* **1971**, *323*, 321–340.
20. Aleshin, V.; Matar, O.B.; van den Abeele, K. Method of memory diagrams for mechanical frictional contacts subject to arbitrary 2D loading. *Int. J. Solids Struct.* **2015**, *60*, 84–95. [CrossRef]
21. Aleshin, V.; Delrue, S.; Trifonov, A.; Matar, O.B.; van den Abeele, K. Two dimensional modeling of elastic wave propagation in solidscontaining cracks with rough surfaces and friction—Part I: Theoretical Background. *Ultrasonics* **2018**, *82*, 11–18. [CrossRef]
22. Aleshin, V.; Delrue, S.; Trifonov, A.; Matar, O.B.; van den Abeele, K. Two dimensional modeling of elastic wave propagation in solidscontaining cracks with rough surfaces and friction—Part II: Numerical implementation. *Ultrasonics* **2018**, *82*, 19–30. [CrossRef]
23. Biwa, S. Second-Harmonic Generation at Contacting Interfaces. In *Nonlinear Ultrasonic and Vibro-Acoustical Techniques for Nondestructive Evaluation*; Springer: Cham, Germany, 2019; pp. 263–299.
24. Saidoun; Meziane, A.; Rénier, M.; Bacon, C.; Zhang, F.; Walaszek, H. Influence of contact interface morphology on the nonlinear interaction between a longitudinal wave and a contact interface with friction: A numerical study, submitted to wave motion. Article submitted for publication.
25. Van den Abeele, K.A.; Carmeliet, J.; Cate, J.A.T.; Johnson, P.A. Nonlinear elastic wave spectroscopy (NEWS) techniques to discern material damage, Part II: Single-mode nonlinear resonance acoustic spectroscopy. *J. Res. Nondestruct. Eval.* **2000**, *12*, 31–42. [CrossRef]

26. Gonzalez-Valadez, M.; Baltazar, A.; Dwyer-Joyce, R. Study of interfacial stiffness ratio of a rough surface in contact using a spring model. *Wear* **2010**, *268*, 373–379. [CrossRef]
27. Teufelsbauer, H.; Wang, Y.; Chiou, M.; Wu, W. Flow–obstacle interaction in rapid granular avalanches: DEM simulation and comparison with experiment. *Granul. Matter* **2009**, *11*, 209–220. [CrossRef]
28. Blanloeuil, P.; Croxford, A.J.; Meziane, A. Numerical and experimental study of the nonlinear interaction between a shear wave and a frictional interface. *J. Acoust. Soc. Am.* **2014**, *135*, 1709–1716. [CrossRef]
29. Bossy, M.; Gobet, E.; Talay, D. A symmetrized Euler scheme for an efficient approximation of reflected diffusions. *J. Appl. Probab.* **2004**, *41*, 877–889. [CrossRef]
30. Baltazar, A.; Rokhlin, S.I.; Pecorari, C. On the Relationship Between Ultrasonic and Micromechanical Properties of Contacting Rough Surfaces. *J. Mech. Phys. Solids* **2002**, *50*, 1397–1416. [CrossRef]
31. Massi, F.; Berthier, Y. Contact surface topography and system dynamics of brake squeal. *Wear* **2008**, *265*, 1784–1792. [CrossRef]
32. Krolikowski, J.; Szczepek, J. Prediction of Contact Parameters using Ultrasonic Method. *Wear* **1991**, *148*, 181–195. [CrossRef]
33. Asma, F.; Bouazzouni, A. Finite element model updating using FRF measurements. *Shock Vib.* **2005**, *12*, 377–388. [CrossRef]
34. Ciavarella, M.; Papangelo, A. On the sensitivity of adhesion between rough surfaces to asperity height distribution. *Phys. Mesomech.* **2018**, *21*, 59–66. [CrossRef]
35. Papangelo, A.; Hoffmann, N.; Ciavarella, M. Load-separation curves for the contact of self-affine rough surfaces. *Sci. Rep.* **2017**, *7*, 6900.

lubricants

MDPI

Article

Friction-Induced Vibration Suppression via the Tuned Mass Damper: Optimal Tuning Strategy

Jia Lin Hu [1] and Giuseppe Habib [2,*]

[1] Department of Applied Mechanics, Budapest University of Technology and Economics, 1111 Budapest, Hungary; hujialin1995@gmail.com

[2] Department of Applied Mechanics, MTA-BME Lendület Machine Tool Vibration Research Group, Budapest University of Technology and Economics, 1111 Budapest, Hungary

* Correspondence: habib@mm.bme.hu

Received: 29 October 2020; Accepted: 16 November 2020; Published: 20 November 2020

check for updates

Abstract: Friction-induced vibrations are a significant problem in various engineering applications, while dynamic vibration absorbers are an economical and effective tool for suppressing various kinds of vibrations. In this study, the archetypal mass-on-moving-belt model with an attached dynamic vibration absorber was considered. By adopting an analytical procedure, the optimal tuning of the absorber's parameters was defined. Furthermore, the bifurcations occurring at the loss of stability were analytically investigated; this analysis illustrated that a properly chosen nonlinearity in the absorber's stiffness permits controlling the supercritical or subcritical character of the bifurcation. However, a numerical analysis of the system's dynamics, despite confirming the analytical results, also illustrated that the system's global behavior is only slightly affected by the bifurcation character. Indeed, a dynamic vibration absorber possessing a perfectly linear restoring force function seems to provide the optimal performance; namely, it minimizes the velocity range for which stick–slip oscillations exists.

Keywords: friction-induced vibrations; mass-on-moving-belt; dynamic vibration absorber; tuned mass damper; passive vibrations mitigation

1. Introduction

Friction-induced vibrations (FIVs) are a peculiar type of oscillations generated by the friction acting between two bodies in relative motion. They consist of either the successions of stick and slip phases between the two bodies [1], or of quasi-harmonic oscillations having an approximately sinusoidal displacement–time relation [2,3]. Although for some specific applications these kinds of vibrations are intentionally generated, such in the case of violin strings [4] and singing wine glasses [5], typically they are seen as a detrimental phenomenon, as in the cases of brake squeal [6] or earthquakes [7,8].

Several methods exist for suppressing FIVs. One possibility is to reduce the friction force at the interface utilizing a lubricant. This method is efficient if friction is not required for the device to operate, such as in the case of hinge squeaking; however, it cannot be adopted for brake squeal mitigation, where high friction is strictly required. Most brake squeal suppression methods consist of increasing the system damping, which is obtained with various techniques [6,9]. Experimental observations also illustrated that isolating the natural frequencies of the brake system's components at low frequencies tends to reduce the occurrence of audible brake squeal [10]; however, in many cases, this strategy is not effective [11]. For active methods for suppressing FIV, Cunefare and Graf [12] proposed adopting a dither exciting the system at non-audible frequencies, which can suppress brake squeal. Papangelo and Ciavarella [13] proposed to mitigate FIVs by normal load variation, for which they provided a closed-form solution.

The dynamic vibration absorber (DVA) is a practical tool for suppressing undesired vibrations in several engineering applications. Its classical design [14] consists of a mass attached to the host structure through a spring and a damper. By tuning its natural frequency in correspondence of the frequency to be damped, it is able to dynamically interact with the host structure dissipating vibration energy. It is successfully employed in several engineering fields for the suppression of various kinds of vibrations, such as flutter instabilities [15,16], rolling motions in ships [17], helicopter rotor oscillations [18] and machine tool vibrations [19,20]. Although DVAs are a mature technology, which was first proposed more than one hundred years ago [17], to the authors' knowledge, there are only a few and relatively recent studies addressing its implementation to suppress FIVs. Popp and Rudolph [21] numerically and experimentally analyzed the performance of a DVA for FIV suppression; by utilizing a single-degree-of-freedom (DoF) primary system, they illustrated its beneficial effect. Chatterjee [22] studied the stability properties of an undamped DVA attached to a two-DoF primary system. Very recently, Niknam and Farhang [23] proposed a study similar to that of Chatterjee [22], where they also provided some numerical simulations of the full system, missing in [22]. Despite the promising results obtained in [21–23], a clear tuning strategy of the DVA's parameters for maximizing its performance is still missing. This paper aims to fill this gap by providing a precise tuning of the absorber parameters for optimizing stability properties and studying the behavior of the host system with the attached DVA while stability is lost.

The rest of the paper is organized as follows. In Section 2, the mechanical model, consisting of the host mass-on-moving-belt system and the attached DVA, is introduced. In Section 3, the stability analysis of the host system, without and with the DVA, is performed, providing explicit equations for the optimal tuning of the absorber parameters. In Sections 4 and 5, the bifurcations occurring at the loss of stability of the host system, without and with absorber, are analytically studied. Furthermore, the effect of the addition of a cubic term in the absorber's restoring force is analytically investigated; analytical results are integrated by numerical simulation, illustrating the system's behavior at high amplitudes. In Section 6, conclusions about the benefits and limitations of the DVA are presented.

2. Mechanical Model

2.1. Primary System

The primary system considered in this study is the classical mass-on-moving-belt model, which is an archetypal system for studying FIVs [1,4,21]. As shown in Figure 1, this single-DoF system consists of a mass m_1, a linear spring k_1 and a linear damper c_1. The mass of the system is in contact with the belt, which moves at a constant driving speed v, while the friction coefficient $\mu(v_{\mathrm{rel}})$ of the contact is a function of the relative velocity $v_{\mathrm{rel}} = v - \dot{x}_1$.

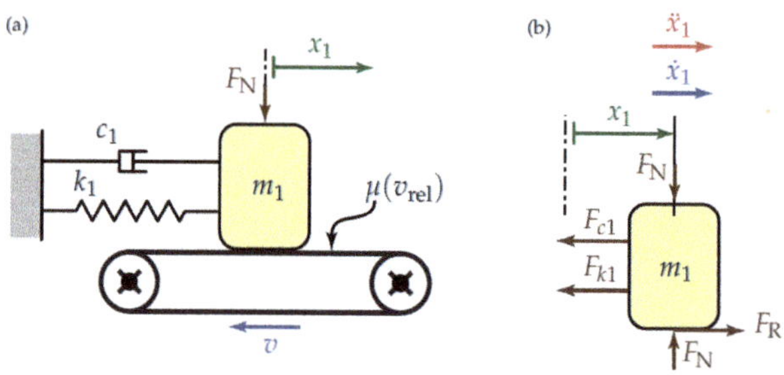

Figure 1. (**a**) The host system without absorber; and (**b**) free body diagram of the host system.

Figure 1b illustrates the free body diagram (FBD) of the host system. The forces acting upon the lumped mass m_1 are the normal forces F_N that cancel each other; the damping and spring forces F_{c1} and F_{k1}, respectively; and the friction force F_R. (For the represented FBD, it is assumed that $v_{\mathrm{rel}} \geq 0$.) Based on Newton's second law of motion and considering that $F_R = \mu\,(v\text{rel})\,F_N$, $F_{c1} = c_1\dot{x}_1$ and $F_{k1} = k_1 x_1$, we obtain that the second order differential equation describing the motion of the one DoF system is

$$m_1\ddot{x}_1 + c_1\dot{x}_1 + k_1 x_1 = F_R\,, \tag{1}$$

with

$$\begin{cases} F_R = \mu\,(v_{\text{rel}})\,F_N & v_{\text{rel}} \neq 0 \\ |F_R| \leq \mu_s F_N & v_{\text{rel}} = 0\,, \end{cases} \tag{2}$$

where the overdot indicates derivation with respect to the time t. The system is in stick condition when the relative velocity is zero ($v_{\text{rel}} = 0$). In this case, the friction force is smaller or equal to $\mu_s F_N$, where μ_s is the static friction. In sliding condition ($v_{\text{rel}} \neq 0$), the direction of the friction force F_R depends on the sign of the relative velocity v_{rel}, which is included in the mathematical formulation of $\mu\,(v_{\text{rel}})$. Additional details about the friction coefficient utilized are provided in Section 2.3. Let us introduce the following expressions

$$\zeta_1 = \frac{c_1}{2\sqrt{m_1 k_1}}\,;\ \omega_{n1} = \sqrt{\frac{k_1}{m_1}}\,;\ x_0 = \frac{F_N}{k_1}\,;\ \tau = \omega_{n1} t\,. \tag{3}$$

By dividing Equation (1) by the mass m_1, applying the expressions from Equation (3) and dividing it by ω_{n1}^2, we obtain

$$x_1'' + 2\zeta_1 x_1' + x_1 = \frac{F_R}{k_1}\,, \tag{4}$$

where prime $'$ indicates derivation with respect to the dimensionless time τ. Then, introducing the dimensionless displacement $\tilde{x}_1 = x_1/x_0$, the system is eventually reduced to

$$\tilde{x}_1'' + 2\zeta_1\tilde{x}_1' + \tilde{x}_1 = \tilde{F}_R\,, \tag{5}$$

where $\tilde{F}_R = F_R/F_N$.

2.2. Mechanical Model of the Host System with the Absorber

We now attach a DVA to the host system. The basis of the model is the same as the one mentioned in the previous subsection. The only additional element is the absorber mass m_2, which is attached to the primary system through a spring and a linear damper. The schematic depiction of this two-DoF system is provided in Figure 2.

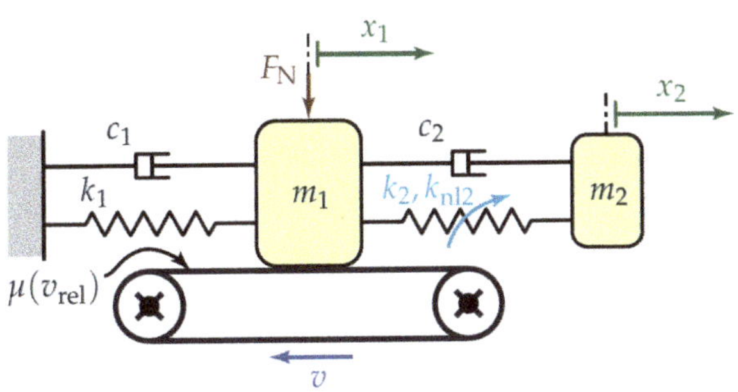

Figure 2. The host system with the attached DVA.

The absorber's spring encompasses a linear term and a cubic term. The differential equations describing the dynamics of the system are

$$
\begin{aligned}
m_1\ddot{x}_1 + c_1\dot{x}_1 + k_1x_1 + c_2(\dot{x}_1 - \dot{x}_2) + k_2(x_1 - x_2) + k_{\text{nl2}}(x_1 - x_2)^3 &= F_{\text{R}} \\
m_2\ddot{x}_2 + c_2(\dot{x}_2 - \dot{x}_1) + k_2(x_2 - x_1) + k_{\text{nl2}}(x_2 - x_1)^3 &= 0\,,
\end{aligned}
\tag{6}
$$

where k_2 and k_{nl2} are the linear and cubic coefficients of the absorber stiffness, respectively, while c_2 is the linear coefficient of the absorber damping. Let us introduce the following expressions

$$
\zeta_2 = \frac{c_2}{2\sqrt{m_2k_2}};\ \omega_{\text{n2}} = \sqrt{\frac{k_2}{m_2}};\ \varepsilon = \frac{m_2}{m_1};\ \gamma = \frac{\omega_{\text{n2}}}{\omega_{\text{n1}}}. \tag{7}
$$

By diving Equation (6) by m_1, applying the expressions from (7) and (3), dividing by ω_{n1}^2 and utilizing the dimensionless time τ and dimensionless displacements $\tilde{x}_1$ and $\tilde{x}_2 = x_2/x_0$, we obtain

$$
\begin{aligned}
\tilde{x}_1'' + 2\zeta_1\tilde{x}_1' + \tilde{x}_1 + 2\varepsilon\zeta_2\gamma(\tilde{x}_1' - \tilde{x}_2') + \varepsilon\gamma^2(\tilde{x}_1 - \tilde{x}_2) + \varepsilon\kappa_{\text{nl2}}(\tilde{x}_1 - \tilde{x}_2)^3 &= \tilde{F}_{\text{R}} \\
\varepsilon\tilde{x}_2'' + 2\varepsilon\zeta_2\gamma(\tilde{x}_2' - \tilde{x}_1') + \varepsilon\gamma^2(\tilde{x}_2 - \tilde{x}_1) + \varepsilon\kappa_{\text{nl2}}(\tilde{x}_2 - \tilde{x}_1)^3 &= 0\,,
\end{aligned}
\tag{8}
$$

where $\kappa_{\text{nl2}} = k_{\text{nl2}}x_0^2/(k_1\varepsilon)$. A variable change is performed, where $\tilde{x}_3 = \tilde{x}_1 - \tilde{x}_2$ (relative displacement of m_2), hence Equation (8) transforms into

$$
\begin{aligned}
\tilde{x}_1'' + 2\zeta_1\tilde{x}_1' + \tilde{x}_1 + 2\varepsilon\zeta_2\gamma\tilde{x}_3' + \varepsilon\gamma^2\tilde{x}_3 + \varepsilon\kappa_{\text{nl2}}\tilde{x}_3^3 &= \tilde{F}_{\text{R}} \\
\varepsilon\left((\tilde{x}_1'' - \tilde{x}_3'') - 2\zeta_2\gamma\tilde{x}_3' - \gamma^2\tilde{x}_3 - \kappa_{\text{nl2}}\tilde{x}_3^3\right) &= 0\,.
\end{aligned}
\tag{9}
$$

2.3. Friction Force

The applied friction law is described by an exponential decaying function, as done for instance in [1], i.e.,

$$
\mu(v_{\text{rel}}) = \left(\mu_{\text{d}} + (\mu_{\text{s}} - \mu_{\text{d}})\,\mathrm{e}^{-\frac{|v_{\text{rel}}|}{\tilde{v}_0}}\right)\operatorname{sgn}(v_{\text{rel}})\,, \tag{10}
$$

where the relative velocity is $v_{\text{rel}} = v - \tilde{x}_1'$. The values assumed by μ for a range of relative velocities for $\mu_{\text{s}} = 1$, $\mu_{\text{d}} = 0.5$ and $v_0 = 0.5$ are represented in Figure 3. The values for the friction law adopted in the present study are the same utilized in [1]. Nevertheless, as illustrated below, the optimization of the absorber parameters does not strictly depend on the considered friction law.

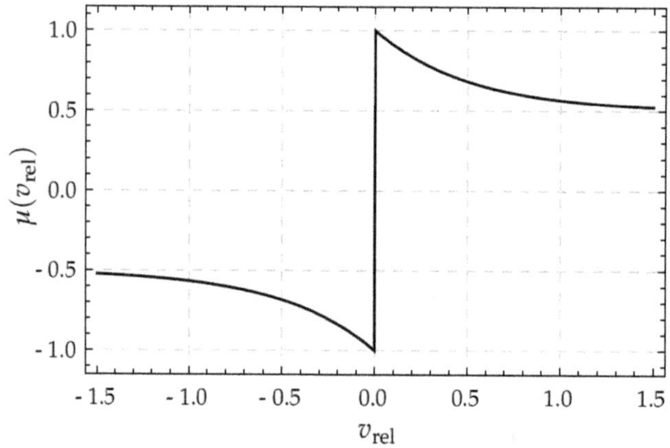

Figure 3. The weakening friction law with $\mu_{\text{s}} = 1$, $\mu_{\text{d}} = 0.5$, $v_0 = 0.5$.

3. Linear Stability Analysis

We now aim at analyzing the system's behavior for small perturbations around the equilibrium. Thus, we linearize it around its trivial solution and we study the stability of this trivial solution. The analysis is performed for both the systems, with and without DVA, in order to assess the beneficial effects of the DVA.

3.1. Linear Stability of the Host System without the DVA

We first linearize Equation (5) around its equilibrium point $x_{1e} = \mu(v_{rel} = v)$, obtaining

$$z_1'' - 2\psi z_1' + z_1 = 0, \tag{11}$$

where $z_1 = x_{1e} + \tilde{x}_1$ and $2\psi = \partial\mu/\partial z_1'|_{z_1'=0} - 2\zeta_1$. Considering the friction law adopted, we have that

$$\left.\frac{\partial\mu}{\partial z_1'}\right|_{z_1'=0} = \frac{\mu_s - \mu_d}{v_0} e^{-\frac{v}{v_0}}, \tag{12}$$

(valid for $v > 0$), which is the slope of the friction force coefficient at the belt velocity v.

Equation (11) corresponds to a linear oscillator, whose trivial solution is asymptotically stable if and only if $\psi < 0$. According to the friction law utilized, and considering that ψ is a monotonically decreasing function of v, the trivial solution is stable for

$$v > v_{h,cr} = v_0 \ln\left(\frac{\mu_s - \mu_d}{2v_0\zeta_1}\right). \tag{13}$$

This result is well-known and better discussed, for instance, in [1]. The practical consequences are that, if the belt moves at a speed lower than $v_{h,cr}$, the equilibrium of the system is unstable and stick–slip oscillations occur. More details about these stick–slip oscillations are provided below.

3.2. Linear Stability of the Host System with DVA

To study the stability of the system with the DVA, we linearize Equation (9) around the equilibrium $\tilde{x}_1 = x_{1e}$ and $\tilde{x}_3 = 0$. By reformulating Equation (9) in explicit form with respect to $\tilde{x}_1''$ and $\tilde{x}_3''$ and by utilizing the variables and parameters introduced in the previous subsection, we obtain

$$\begin{bmatrix} z_1' \\ z_2' \\ z_3' \\ z_4' \end{bmatrix} = \begin{bmatrix} 0 & 0 & 1 & 0 \\ 0 & 0 & 0 & 1 \\ -1 & -\gamma^2\varepsilon & 2\psi & -2\varepsilon\zeta_2\gamma \\ -1 & -\gamma^2(\varepsilon+1) & 2\psi & -2(\varepsilon+1)\gamma\zeta_2 \end{bmatrix} \begin{bmatrix} z_1 \\ z_2 \\ z_3 \\ z_4 \end{bmatrix} = \boldsymbol{A}\boldsymbol{z}, \tag{14}$$

where $z_1 = x_{1e} + \tilde{x}_1$, $z_2 = x_3$, $z_3 = z_1'$ and $z_4 = z_2'$.

We analyze the characteristic exponents of the system to determine the stability. The characteristic polynomial is

$$p(\lambda) = \det(\boldsymbol{A} - \lambda\boldsymbol{I}) = 0. \tag{15}$$

The characteristic polynomial is in the form of

$$p(\lambda) = a_0\lambda^4 + a_1\lambda^3 + a_2\lambda^2 + a_3\lambda + a_4 = 0 \quad (a_0 > 0). \tag{16}$$

The stability is determined based on the *Liénard–Chipart conditions* (LCC) [24]. For the polynomial $p(\lambda)$ to have all roots with negative real parts, it is necessary and sufficient that the coefficients of the polynomial are all positive ($a_i > 0, i = 1, \ldots, 4$) and that the determinant inequalities $\Delta_1 > 0$ and $\Delta_3 > 0$ (Hurwitz determinantal inequalities) are verified, which in our case means

$$a_0 = 1 > 0, \tag{17}$$

$$a_1 = 2\left(\gamma\left(1+\varepsilon\right)\zeta_2 - \psi\right) > 0 \Longleftrightarrow \psi < \left(1+\varepsilon\right)\gamma\zeta_2, \tag{18}$$

$$a_2 = 1 + \gamma\left(\gamma\left(1+\varepsilon\right) - 4\zeta_2\psi\right) > 0 \Longleftrightarrow \psi < \frac{1+\gamma^2\left(1+\varepsilon\right)}{4\gamma\zeta_2}, \tag{19}$$

$$a_3 = 2\gamma\left(\zeta_2 - \gamma\psi\right) > 0 \Longleftrightarrow \psi < \frac{\zeta_2}{\gamma}, \tag{20}$$

$$a_4 = \gamma^2 > 0, \tag{21}$$

$$\Delta_1 = a_1 = 2\gamma\left(1+\varepsilon\right)\zeta_2 - 2\psi > 0 \text{ (Already present in (18))}, \tag{22}$$

$$\begin{aligned}\Delta_3 = a_1 a_2 a_3 - a_0 a_3^2 - a_1^2 a_4 = -4\gamma\Big(&\gamma^4(\varepsilon+1)^2\zeta_2\psi - \gamma^3\psi^2\left(4(\varepsilon+1)\zeta_2^2 + \varepsilon\right) \\ &+ 2\gamma^2\zeta_2\psi\left(2(\varepsilon+1)\zeta_2^2 + 2\psi^2 - 1\right) - \gamma\zeta_2^2\left(\varepsilon + 4\psi^2\right) + \zeta_2\psi\Big) > 0.\end{aligned} \tag{23}$$

3.2.1. Analytical Optimal Solution

Considering the stability analysis performed in the previous section, we aim at finding the parameter values of the absorber which maximize the stable region. The linear system in Equation (14) has the same mathematical form as the one studied in [25]. Thus, we can follow the same steps in order to optimize the absorber.

First, we look at the curves where the coefficients and the Hurwitz determinants are zeros; these are the boundaries where certain roots change. At certain boundaries, the stable/unstable transition takes place. Following the steps discussed in [25], we can define specific points on these curves, which helps us find the optimal parameters. Figure 4 shows the stability regions for different values of ζ_2; the boundary curves of the LCC are also depicted. The gray shaded region is the stable region; we can observe that the stability boundary is not at a constant value of ψ as it is the case for the host system without the DVA; instead, it is a function of γ, with a pronounced peak for $\gamma \approx 1$. This is an expected feature, considering that the DVA usually needs to be tuned at a frequency close to the natural frequency of the primary system [14]. For different values of ζ_2, the maximum value of ψ also changes, thus we need to find the optimal combination of (ζ_2, γ) such that the value of ψ generating instabilities is maximized. For the optimization, the mass ratio ε is assumed constant; however, the results of the analysis show that larger values of ε increase the stable region.

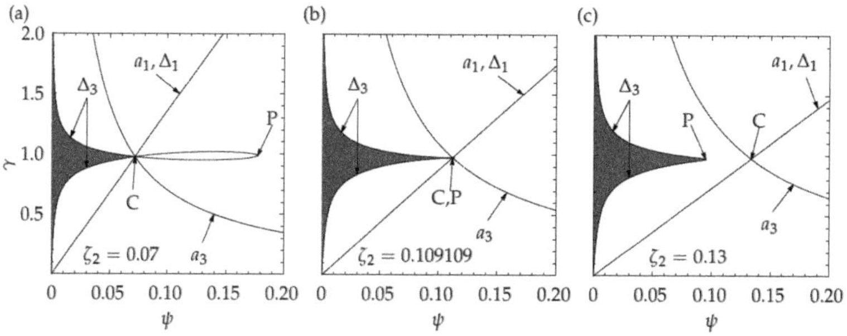

Figure 4. Stability diagrams for different values of ζ_2 and $\varepsilon = 0.05$: (**a**) $\zeta_2 = 0.07$; (**b**) $\zeta_2 = 1/2\sqrt{\varepsilon/\left(1+\varepsilon\right)} = 0.109109$; and (**c**) $\zeta_2 = 0.13$.

As we can see in Figure 4a, the intersection of the curves $a_1 = \Delta_1 = 0$ and $a_3 = 0$ defines a point that we denote with C. Its coordinates in the (ψ, γ) plane are $\text{C} = \left(\zeta_2\sqrt{1+\varepsilon}, 1/\sqrt{1+\varepsilon}\right)$. For low values of ζ_2, Point C marks the maximal value of ψ providing stability, which we call ψ^*. Increasing the value of ζ_2, Figure 4c illustrates that Point C does not identify any more ψ^* and that the stability boundary is defined by Equation $\Delta_3 = 0$. Nevertheless, also for large damping ζ_2, the value of γ which maximizes ψ^* is very close to the γ coordinate of Point C. Therefore, we define Point P as the point of the curve $\Delta_3 = 0$ having the same γ coordinate as C. Substituting the γ coordinate of C, $\gamma = 1/\sqrt{1+\varepsilon}$, into $\Delta_3 = 0$, we obtain Point P $= \left(\varepsilon/\left(4\zeta_2\sqrt{1+\varepsilon}\right), 1/\sqrt{1+\varepsilon}\right)$.

Since Points C and (approximately) P alternatively mark ψ^*, depending on the value of the absorber damping ζ_2, by choosing ζ_2 such that P and C coincide, we can maximize ψ^*. Imposing equality between the ψ coordinates of C and P, we attain

$$\zeta_2\sqrt{1+\varepsilon} = \frac{\varepsilon}{\left(4\sqrt{1+\varepsilon}\zeta_2\right)}, \tag{24}$$

therefore, ψ^* is maximized for

$$\gamma = \gamma_{\text{opt}} = \frac{1}{\sqrt{1+\varepsilon}} \quad \text{and} \quad \zeta_2 = \zeta_{2\text{opt}} = \frac{1}{2}\sqrt{\frac{\varepsilon}{1+\varepsilon}}, \tag{25}$$

and the corresponding maximal value of ψ^* is

$$\psi_{\max} = \frac{\sqrt{\varepsilon}}{2}. \tag{26}$$

Accordingly, if $\gamma = \gamma_{\text{opt}}$ and $\zeta_2 = \zeta_{2\text{opt}}$, the system is stable if $2\psi < \sqrt{\varepsilon}$, or, in other words, if

$$\frac{\mu_s - \mu_d}{v_0} e^{-\frac{v}{v_0}} - 2\zeta_1 < \sqrt{\varepsilon}. \tag{27}$$

Considering that ψ is a monotonically decreasing function of v, the equilibrium of the system is stable if

$$v > v_{cr} = v_0 \log\left(\frac{\mu_s - \mu_d}{v_0\left(\sqrt{\varepsilon} + 2\zeta_1\right)}\right). \tag{28}$$

The stability chart corresponding to this case is illustrated in Figure 4b. We remark that the optimal tuning proposed here does not strictly depend on the friction law considered, which could be modeled with alternative functions without varying γ_{opt} and $\zeta_{2\text{opt}}$. This represent a clear advantage in the case of real engineering applications.

3.2.2. Numerical Validation

The optimization procedure utilized in the previous section is based on a heuristic approach, which does not prove that γ_{opt} and $\zeta_{2\text{opt}}$ provide the maximal possible value of ψ^*. Therefore, its validity should be verified numerically. The numerical analysis is performed by directly computing eigenvalues of matrix A on a grid of the (ζ_2, γ, ψ) space for a fixed ε value ($\varepsilon = 0.05$) and identifying the couple of values ζ_2, γ which provides the maximal ψ^*. After several trials, the analysis is finally performed on the grid described in Table 1.

Table 1. Parameter grid for the optimum search.

Parameter	min	max	Step
ζ_2	0.1	0.11	10^{-5}
γ	0.97589	0.97591	10^{-6}
ψ	0.1124	0.1126	10^{-5}

This analysis provided the optimal values for ζ_2 and γ, which are indicated in Table 2 and directly compared with the optimal values obtained analytically.

Table 2. Comparison of numerical and analytical results.

Parameter	Numerical	Analytical	Relative Error [%]
ζ_2	0.10977	0.109109	0.605867 [%]
γ	0.975899	0.975900	0.000109945 [%]
ψ	0.11247	0.111803	0.596226 [%]
v_{cr}	1.36678	1.36883	0.364409 [%]

Although numerical and analytical optimal parameters do not coincide, their difference is minimal and negligible for most engineering applications. In particular, the optimal γ value is practically the same in both cases. The critical velocity, computed utilizing the parameter values indicated in Table 3, has a difference of less than 0.4% in the two cases. We remark that the difference between numerical and analytical computation is not related to the inaccuracy of stability estimation through the LCC, which exactly predicts an equilibrium's stability, but to the heuristic approach utilized for the optimization.

Table 3. Numerical input data.

μ_s	μ_d	v_0	ε	ζ_1
1	0.5	0.5	0.05	0.05

Figure 5, illustrating the curve $\Delta_3 = 0$ for various values of ζ_2, enables us to understand the reason for the difference between the results obtained with the numerical and analytical approach. The blue line in the figure corresponds to the analytical optimization, while the green line to the numerical one. The yellow and red curves refer to values of ζ_2 slightly higher and lower than the optimal ones, respectively. The inaccuracy of the analytical procedure is due to the fact that the peak of the $\Delta_3 = 0$ curve does not exactly lie on Point P (which has a fixed γ value and it is not represented in the figure). However, considering the minimal difference found and the practical compactness of Equations (24) and (25), these will be utilized in the continuation of the paper.

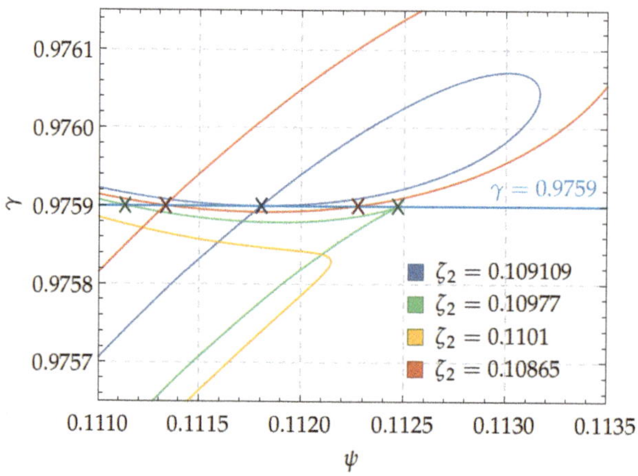

Figure 5. Vanishing loop of the $\Delta_3 = 0$ curve. The blue line was obtained utilizing the optimal damping as defined by the analytical procedure $\zeta_2 = \zeta_{2opt}$, the green line corresponds to the optimal solution obtained by the numerical procedure and the yellow and red lines are obtained for ζ_2 values slightly larger and smaller, respectively, than ζ_{2opt}.

3.3. Evaluation of the Absorber's Performance

We now aim at evaluating the performance of the DVA by performing various comparison between the system with and without DVA. As mentioned above, the critical velocities for both cases are

$$v_{\mathrm{h,cr}} = v_0 \ln\left(\frac{\mu_\mathrm{s} - \mu_\mathrm{d}}{2v_0\zeta_1}\right), \tag{29}$$

$$v_{\mathrm{cr}} = v_0 \ln\left(\frac{\mu_\mathrm{s} - \mu_\mathrm{d}}{v_0\left(\sqrt{\varepsilon} + 2\zeta_1\right)}\right). \tag{30}$$

Considering the characteristics of a logarithm function, we can state that, if the argument becomes 1, the logarithm function yields 0. Thus, there is a certain parameter set for which the critical velocity becomes 0 (inherent stability). We define $\zeta_{1,\mathrm{cr}}$ as the critical primary damping parameter, for which the critical velocity becomes 0. To obtain $\zeta_{1,\mathrm{cr}}$, we solve the arguments of the logarithm functions for 1; these yield

$$\frac{\mu_\mathrm{s} - \mu_\mathrm{d}}{2v_0\zeta_{1\mathrm{h,cr}}} = 1 \longrightarrow \zeta_{1\mathrm{h,cr}} = \frac{\mu_\mathrm{s} - \mu_\mathrm{d}}{2v_0}, \tag{31}$$

$$\frac{\mu_\mathrm{s} - \mu_\mathrm{d}}{v_0\left(\sqrt{\varepsilon} + 2\zeta_{1,\mathrm{cr}}\right)} = 1 \longrightarrow \zeta_{1,\mathrm{cr}} = \frac{\mu_\mathrm{s} - \mu_\mathrm{d} - v_0\sqrt{\varepsilon}}{2v_0}. \tag{32}$$

Utilizing values in Table 3, the numerical values for the critical primary damping are

$$\zeta_{1\mathrm{h,cr}} = 0.5\,, \tag{33}$$

$$\zeta_{1,\mathrm{cr}} = 0.388197\,. \tag{34}$$

This shows that the application of an optimally tuned DVA with a mass of only 5% of the host system mass enables to reduce the critical primary damping of 22%.

Considering, instead, the critical velocity as a base of comparison, we define the improvement factor φ such that

$$\varphi := \frac{\tilde{v}_{\mathrm{h,cr}} - \tilde{v}_{\mathrm{cr}}}{\tilde{v}_{\mathrm{h,cr}}} \times 100\%\,. \tag{35}$$

The critical velocities for both cases and the improvement factor are plotted against the varying ζ_1 in Figure 6, utilizing the parameter values in Table 3. We can observe that the difference in critical velocity is more significant for smaller values of ζ_1, i.e., for a slightly damped host system. We also notice that, if the host system is completely undamped, then the critical velocity is undefined, meaning that the equilibrium is always unstable. For any value of the host system damping ζ_1, the improvement factor is almost always above 50%.

Let us also observe what happens if we vary the value of the mass ratio ε. Similar to before, the critical velocities for both systems and the improvement factor curve are illustrated in Figure 7. The parameter values are those indicated in Table 3. The critical velocity of the host system is constant because it does not depend on ε; however, for the system with the DVA, the critical velocity monotonously decreases with ε. Utilizing the parameter values in Table 3, the critical velocities for both the host system and the system with DVA are $v_{\mathrm{h,cr}} = 1.151$ and $v_{\mathrm{cr}} = 0.5641$, hence the improvement provided by the DVA is of $\varphi = 51\%$.

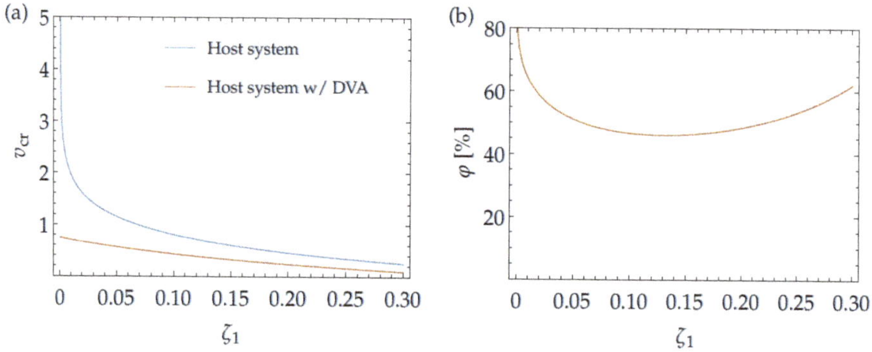

Figure 6. Comparison of host system with and without DVA with varying ζ_1, other parameters as in Table 3: (**a**) critical velocities; and (**b**) improvement curve.

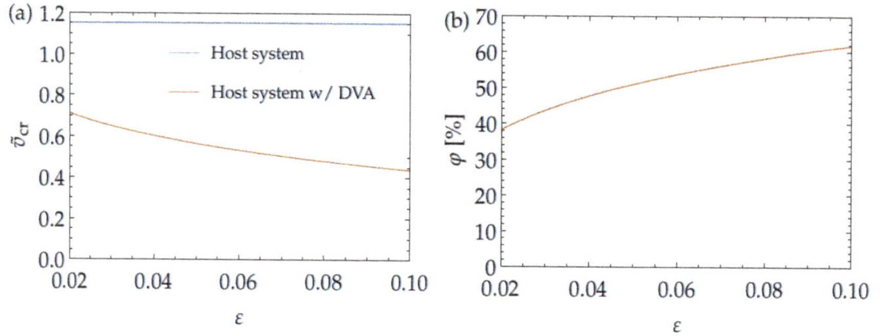

Figure 7. Comparison of host system with and without DVA with varying ε: (**a**) critical velocities; and (**b**) improvement curve.

4. Bifurcation Analysis of the Host System without the DVA

The analysis performed in Section 3 refers to the system linearized around its equilibrium. Therefore, it is able to describe its dynamics only in the vicinity of the equilibrium, while phenomena occurring when the stability is lost are overlooked. Additionally, it provides no information about the stable equilibrium's robustness if the system is subject to non-small perturbations. To investigate the system behavior at the loss of stability and correctly evaluate the DVA performance, we reintroduce the nonlinear terms and analytically perform a bifurcation analysis of the system without and with DVA.

Considering the system in Equation (5), we first center the system around its equilibrium point x_{1e} by introducing the variable $z_1 = x_{1e} + \tilde{x}_1$, and then we expand it in Taylor series up to the third order, obtaining

$$\begin{bmatrix} z_1' \\ z_3' \end{bmatrix} = \begin{bmatrix} 0 & 1 \\ -1 & \frac{\mu_d-\mu_s}{v_0}\mathrm{e}^{-\frac{v}{v_0}} - 2\zeta_1 \end{bmatrix} \begin{bmatrix} z_1 \\ z_3 \end{bmatrix} + \begin{bmatrix} 0 \\ -\left(\frac{\mu_d-\mu_s}{2v_0^2}\mathrm{e}^{-\frac{v}{v_0}}\right) z_3^2 - \left(\frac{\mu_d-\mu_s}{6v_0^3}\mathrm{e}^{-\frac{v}{v_0}}\right) z_3^3 \end{bmatrix} = \boldsymbol{A}_{\mathrm{h}}\mathbf{z}_{\mathrm{h}} + \boldsymbol{b}_{\mathrm{h}} \,. \tag{36}$$

For $v \approx v_{\mathrm{h,cr}}$, matrix $\boldsymbol{A}_{\mathrm{h}}$ has complex conjugate eigenvalues $\lambda_{1,2\mathrm{h}} = \alpha_{1\mathrm{h}} \pm \mathrm{i}\omega_{1\mathrm{h}}$ and eigenvectors $\mathbf{s}_1 = \bar{\mathbf{s}}_2$, which are reduced to

$$\lambda_{1\mathrm{h}} = \mathrm{i},\ \lambda_{2\mathrm{h}} = -\mathrm{i},\ \mathbf{s}_1 = \begin{bmatrix} -\mathrm{i} \\ 1 \end{bmatrix},\ \mathbf{s}_2 = \begin{bmatrix} \mathrm{i} \\ 1 \end{bmatrix} \tag{37}$$

for $v = v_{h,cr}$.

We then define the transformation matrix

$$T_h = \left[\mathrm{Re}\,(s_2) \quad \mathrm{Im}\,(s_2)\right]\Big|_{v=v_{h,cr}}, \tag{38}$$

we apply the transformation $z = T_h y_h$ and we pre-multiply Equation (36) by T_h^{-1}, leading to

$$y_h' = T_h^{-1} A_h T_h y_h + T_h^{-1} b_h = W_h y_h + \tilde{b}_h, \tag{39}$$

where

$$W_h = \begin{bmatrix} \alpha_{1h}(v) & -1 \\ 1 & \alpha_{1h}(v) \end{bmatrix} \text{ and } \tilde{b}_h = \begin{bmatrix} -\left(\frac{\mu_d-\mu_s}{2v_0^2} e^{-\frac{v}{v_0}}\right) y_2^2 - \left(\frac{\mu_d-\mu_s}{6v_0^3} e^{-\frac{v}{v_0}}\right) y_2^3 \\ 0 \end{bmatrix}. \tag{40}$$

For $v = v_{h,cr}$, $\alpha_{1h} = 0$, α_{1h} is kept as a generic function of v, since α_{1h} is the critical term causing the instability. The system in Equation (39) is in the so-called Jordan normal form.

By performing several transformations, namely transformation in complex form, near-identity transformation and transformation in polar coordinates, the bifurcation can be characterized through its normal form

$$r' = \alpha_{1h}(v) r + \delta_h r^3, \tag{41}$$

where

$$\delta_h = \frac{e^{-\frac{v}{v_0}} (\mu_s - \mu_d)}{16 v_0^3}. \tag{42}$$

Details of this standard procedure can be found in [26]. Non-zero real equilibrium solutions of Equation (41) correspond to periodic motion of the system in Equation (5). Linearizing $\alpha_{1h}(v)$ around $v = v_{h,cr}$, we obtain

$$r' = r\left(\alpha_{1h}^* (v - v_{h,cr}) + \delta_h r^2\right), \text{ where } \alpha_{1h}^* = \frac{d\alpha_{1h}}{dv}\Big|_{v=v_{h,cr}} = -e^{-\frac{v}{v_0}} \frac{\mu_s - \mu_d}{2v_0^2}, \tag{43}$$

which has solutions

$$r = r_{h0} = 0 \text{ and } r = r_h^* = \sqrt{-\frac{\alpha_{1h}^* (v - v_{h,cr})}{\delta_h}} = 2\sqrt{2v_0 (v - v_{h,cr})}. \tag{44}$$

The trivial solution r_{h0} exists for any value v and it is stable for $v > v_{h,cr}$. Differently, r_h^* is real only if the argument of the square root in Equation (44) is non-negative, which occurs for $v > v_{h,cr}$. Since $\mu_s > \mu_d$, in all relevant cases δ is positive (see Equation (42)), which, as clearly explained in [26], means that the bifurcation is subcritical. This implies that r_h^* corresponds to unstable solutions of Equation (41). This result is in accordance with [1].

A practical consequence of the subcritical character of the bifurcation is that the system, even within the stable region of the equilibrium ($v > v_{h,cr}$), can experience large oscillations. If the system, while in equilibrium, is subject to a sufficiently large perturbation, which makes it cross the unstable periodic solution in the phase space, it will leave its basin of attraction and it will reach another attractor, which in this case consists of stick–slip oscillations.

The bifurcation diagram in Figure 8a clearly illustrates this scenario. The dashed line indicates a branch of unstable periodic solutions generated at the bifurcation (this branch was obtained through time reverse numerical simulations). The solid line, instead, marks the branch of stick–slip oscillation. The thin solid red line represents the branch of unstable periodic solutions obtained from the analytical computation. We remark on the excellent agreement of the analytically computed solution with the numerical one at low amplitudes. For $v \in [1.15, 1.83]$, the system presents two stable solutions, the trivial one and a stick–slip periodic solution, and an unstable periodic solution, as illustrated in Figure 8b for $v = 1.3$. Depending on the initial conditions, the system will either converge towards the trivial solution (red curve in Figure 8c) or will undergo stick–slip oscillations (blue curve in Figure 8c). Numerical solutions were computed utilizing the switch model proposed in [4].

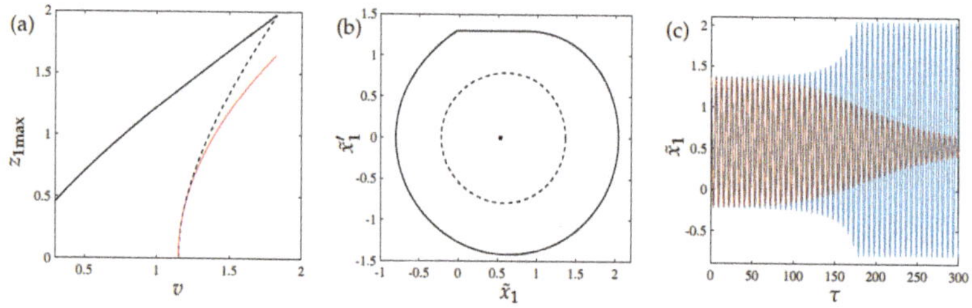

Figure 8. (**a**) Bifurcation diagram for the host system without DVA; the thin red line marks analytical solutions, black lines numerical ones and dashed lines the numerical unstable solutions. (**b**) Steady state solutions of the system for $v = 1.3$; solid line is the stable solution and dashed line is the unstable solution. (**c**) Time series of the system leading to the steady state solutions represented in (**b**) with initial conditions $z_h = [1.367, 0]^{\mathrm{T}}$ (blue line) and $z_h = [1.36, 0]^{\mathrm{T}}$ (red line). Other parameter values are as in Table 3.

5. Bifurcation Analysis of the Host System with the DVA

To evaluate the DVA performance when stability is lost, we analyze the bifurcation behavior of the system with an attached DVA. The analysis is performed assuming that γ and ζ_2 are tuned approximately according to Equation (25). An analysis of the eigenvalues of matrix A illustrates that at the loss of stability, if γ and ζ_2 are tuned approximately according to Equation (25), a couple of complex conjugate eigenvalues leaves the left-hand side of the complex plane, meaning that their real parts become positive. This scenario corresponds to the occurrence of a Hopf bifurcation. We also notice that, if $\zeta_2 \leq \zeta_{2\mathrm{opt}}$ and $\gamma = \gamma_{\mathrm{opt}}$, not one, but two couples of complex conjugate eigenvalues leave the left-hand side of the complex plane. Referring to the stability chart in Figure 4a, the entire unstable region matrix A has only one couple of eigenvalues with positive real part, except in the loop delimited by Points C and P, where all four eigenvalues have positive real part. This scenario corresponds to a Hopf–Hopf (or double Hopf) bifurcation. In the following, the case of a single Hopf bifurcation is analyzed.

The first step of the analysis consists of transforming the system in Equation (9) into first-order form, similar to Equation (14), but including nonlinear terms up to the third order, which leads to

$$\begin{bmatrix} z_1' \\ z_2' \\ z_3' \\ z_4' \end{bmatrix} = \begin{bmatrix} 0 & 0 & 1 & 0 \\ 0 & 0 & 0 & 1 \\ -1 & -\gamma^2\varepsilon & 2\psi & -2\varepsilon\zeta_2\gamma \\ -1 & -\gamma^2(\varepsilon+1) & 2\psi & -2(\varepsilon+1)\gamma\zeta_2 \end{bmatrix} \begin{bmatrix} z_1 \\ z_2 \\ z_3 \\ z_4 \end{bmatrix} + \begin{bmatrix} 0 \\ 0 \\ -\frac{z_3^2(\mu_{\mathrm{d}}-\mu_{\mathrm{s}})e^{-\frac{v}{v_0}}(3v_0+z_3)}{6v_0^3} - \varepsilon k_{\mathrm{nl2}} z_2^3 \\ -\frac{z_3^2(\mu_{\mathrm{d}}-\mu_{\mathrm{s}})e^{-\frac{v}{v_0}}(3v_0+z_3)}{6v_0^3} - (\varepsilon+1)\,\varepsilon k_{\mathrm{nl2}} z_2^3 \end{bmatrix}$$
$$= Az + b\,. \tag{45}$$

In the vicinity of the loss of stability, A has two couples of complex conjugate eigenvalues $\lambda_{1,2} = \alpha_1 \pm \mathrm{i}\omega_1$ and $\lambda_{3,4} = \alpha_2 \pm \mathrm{i}\omega_2$. To decouple the linear part of the system, we define the transformation matrix

$$T = \begin{bmatrix} \mathrm{Re}\,(s_1) & \mathrm{Im}\,(s_1) & \mathrm{Re}\,(s_3) & \mathrm{Im}\,(s_3) \end{bmatrix}, \tag{46}$$

where s_1 and s_3 are two of the eigenvectors of A, and we apply the transformation $z = Ty$, obtaining

$$\dot{y} = T^{-1}Ay + T^{-1}b = Wy + \tilde{b}\,, \tag{47}$$

where

$$\mathbf{W} = \begin{bmatrix} \alpha_1 & \omega_1 & 0 & 0 \\ -\omega_1 & \alpha_1 & 0 & 0 \\ 0 & 0 & \alpha_2 & \omega_2 \\ 0 & 0 & -\omega_2 & \alpha_2 \end{bmatrix}. \quad (48)$$

For the sake of brevity, the explicit formulation of $\tilde{\boldsymbol{b}}$ is omitted here.

In the case of a single Hopf bifurcation, only α_1 becomes positive at the loss of stability, while α_2 remains negative. Therefore, only the first two equations of Equations (47) are linearly related to the bifurcation, while y_3 and y_4 have minor local effect at the bifurcation. Next, we aim at reducing the dynamics of the system to the so-called *center manifold*, which is a two-dimensional surface tangent at the bifurcation point to the subspace spanned by the two eigenvectors s_1 and s_2 related to the bifurcation. To do so, we approximate y_3 and y_4 by $y_3 = \eta_{320}y_1^2 + \eta_{311}y_1y_2 + \eta_{302}y_2^2$ and $y_4 = \eta_{420}y_1^2 + \eta_{411}y_1y_2 + \eta_{402}y_2^2$, reducing the system to

$$\begin{aligned} y_1 &= \alpha_1 y_1 + \omega_1 y_2 + \sum_{j+k=2,3} a_{jk} y_1^j y_2^k + \text{h.o.t.} \\ y_2 &= -\omega_1 y_1 + \alpha_1 y_2 + \sum_{j+k=2,3} b_{jk} y_1^j y_2^k + \text{h.o.t.}, \end{aligned} \quad (49)$$

where j and k are non-negative integers (more details on this procedure can be found in [26]) and h.o.t. stands for higher order terms.

The system in Equation (49) has the same form as Equation (39); therefore, exactly the same steps can be performed to reduce the system to its normal form, that is

$$r' = \alpha_1(v)r + \delta r^3, \quad (50)$$

where [26]

$$\delta = \frac{1}{8}\left(\frac{1}{\omega_1}\left((a_{20} + a_{02})(-a_{11} + b_{20} - b_{02}) + (b_{20} + b_{02})(a_{20} - a_{02} + b_{11})\right) + (3a_{30} + a_{12} + b_{21} + 3b_{03})\right). \quad (51)$$

Imposing $\varepsilon = 0.05$, $\gamma = \gamma_{\text{opt}}$ and $\zeta_2 = 1.05\,\zeta_{2\text{opt}}$, we obtain

$$\delta = 0.00474 + 0.68\,k_{\text{nl2}}\,. \quad (52)$$

Proceeding as done for the host system without DVA, we have that the non-trivial solutions of Equation (50) is given by

$$r = r^* = \sqrt{-\frac{\alpha_1^*(v - v_{\text{cr}})}{\delta}}, \quad (53)$$

where $\alpha_1^* = \mathrm{d}\alpha_1/\mathrm{d}v|_v = v_{\text{cr}}$. We notice that δ is positive if the DVA is linear ($k_{\text{nl2}} = 0$), which means that also in this case the bifurcation is subcritical, and it generates unstable periodic solutions. Analyzing other values of ζ_2 and ε, we verified that the subcritical characteristic persists for a relatively large parameter value range. The corresponding bifurcation diagram is illustrated in Figure 9a. Comparing Figures 9a and 8a, we notice that, although the linear DVA does not change the characteristic of the bifurcation, the advantages in terms of vibrations suppression persist also in the nonlinear range. In fact, for the considered parameter values, in the host system without DVA, stick–slip oscillations exist for $v \in (0, 1.83]$, while, with the addition of the absorber, they are limited to the range $v \in (0, 0.768]$.

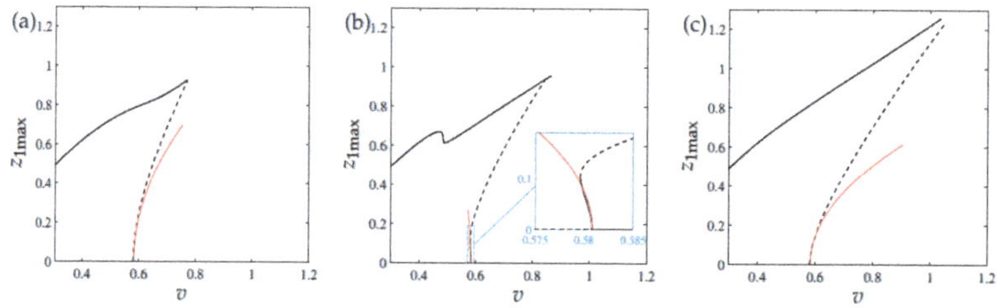

Figure 9. Bifurcation diagrams for the host system with the DVA for parameter values as in Table 3, $\gamma = \gamma_{opt}$ and $\zeta_2 = \zeta_{2opt}$: (**a**) $k_{nl2} = 0$, (**b**) $k_{nl2} = -0.01$; and (**c**) $k_{nl2} = 0.01$. Solid lines are stable solutions, dashed lines are unstable solutions and thin red lines are analytical solutions.

Equation (52) suggests that, if $k_{nl2} < -0.00697$, δ becomes negative, making therefore the bifurcation supercritical. This scenario is confirmed by the bifurcation diagram depicted in Figure 9b, for $k_{nl2} = -0.01$. Although at first sight it seems that the bifurcation is subcritical, the inset illustrates that the bifurcation is indeed supercritical; however, the branch of periodic solutions bends rapidly to the right in correspondence of a fold bifurcation, making the overall scenario similar to the case of $k_{nl2} = 0$. The figure confirms the correctness of the analytical computation; nevertheless, it also points out that the performed local analysis is unable to capture the global behavior of the system, which is not qualitatively affected by the variation of the nonlinear characteristic of the DVA's spring. Furthermore, we notice that the addition of the softening nonlinear spring enlarges the bistable range, making stick–slip oscillations exist up to $v < 0.86$, instead of 0.768 as in the case of $k_{nl2} = 0$.

Figure 9c illustrates the bifurcation diagram obtained for a hardening absorber's spring ($k_{nl2} = 0.01$). In this case, the range of existence of stick–slip oscillations is further enlarged, persisting up to $v < 1.035$. We also remark that increasing the value of k_{nl2} above 0.01 or decreasing it below -0.01 provided only worse performance than those illustrated in Figure 9. This result suggests that any low order nonlinearity of the absorber's stiffness is detrimental concerning the DVA effectiveness. This finding is somehow surprising, considering that in similar applications the addition of a properly tuned nonlinear term in the DVA's stiffness provided some advantages [20,25,27].

Regarding Figure 9b, we notice that the branch of stick–slip oscillations presents two folds for $v \approx 0.48$. However, an analysis of the system's steady state solutions before and after the folds did not reveal any particular detail relevant from an engineering point of view; therefore, the phenomenon was not analyzed in further detail. We also remark that, in Figure 9b,c, the branches of stable and unstable solutions do not encounter each other at a well defined point, as happens in Figure 8a, for instance. This is probably related to the fact that the branches of unstable solutions in Figure 9 were obtained adopting the shooting method (employing `MatCont` [28], a `MATLAB`-based toolbox for numerical continuation) of the system smoothed assuming that v_{rel} is always positive. This assumption makes the considered system unable to exhibit stick–slip oscillations, but keeps it equivalent to the original system for $v > z_3$. In contrast, the stable branches were obtained from direct numerical simulations of the full system. Therefore, inaccuracies of the smoothed system in the proximity of the onset of stick–slip motions are possible.

As mentioned at the beginning of this section, for $\zeta_2 < \zeta_{2opt}$ and $\gamma = \gamma_{opt}$, the system undergoes a Hopf–Hopf bifurcation. However, acknowledging the fact that the bifurcation analysis seems to be an inefficient tool for investigating the post-bifurcation behavior of the system, which is dominated by large amplitude oscillations, and considering that the analysis of such a bifurcation requires a significant analytical effort, the detailed investigation of this case is omitted in this study.

6. Conclusions

In this study, the problem of suppressing FIVs through a DVA was addressed. Possibly the simplest system exhibiting FIVs was considered, i.e., the mass-on-moving-belt system, to which a classical DVA was attached. The optimal tuning of the absorber parameters was defined through an analytical procedure, which enabled us to reduce the critical velocity by approximately 50%, with an additional mass of only 5% of the primary system's mass.

The post-bifurcation behavior analysis illustrated that, although a linear DVA is unable to change the bifurcation character at the loss of stability, it can still significantly reduce the extent of the bistable region. Globally, the area of existence of stick–slip oscillations is reduced by 58%, with a DVA mass of only 5%. The bifurcation analysis proved that it is possible to change the bifurcation character if a small softening term is included in the absorber. However, this has only a local beneficial effect, while, globally, it enlarges the region of existence of stick–slip motions. The performance also worsens if an additional hardening term is introduced, suggesting that the spring characteristic should be maintained as linear as possible. Large order nonlinearities, such as non-smoothness, might have beneficial effects; nevertheless, their analysis was not addressed in this study, and it is left for future developments. Other possible future developments of the present study include the analysis of the Hopf–Hopf bifurcation occurring at the loss of stability for $\zeta_2 < \zeta_{2opt}$ and the analysis of the performance of the DVA if the primary system has two DoF, encompassing, therefore, coupling instabilities as well [22].

Author Contributions: Conceptualization, G.H.; Data curation, J.L.H. and G.H.; Formal analysis, J.L.H. and G.H.; Funding acquisition, G.H.; Investigation, J.L.H. and G.H.; Methodology, J.L.H. and G.H.; Resources, G.H.; Software, J.L.H. and G.H.; Supervision, G.H.; Validation, J.L.H. and G.H.; Visualization, J.L.H. and G.H.; Writing—original draft, J.L.H. and G.H.; and Writing—review and editing, J.L.H. and G.H. All authors have read and agreed to the published version of the manuscript.

Funding: This research was funded by Hungarian National Science Foundation under grant number OTKA 134496 and by the NRDI Fund (TKP2020 IES, Grant No. BME-IE-NAT, and TKP2020 NC, Grant No. BME-NC) based on the charter of bolster issued by the NRDI Office under the auspices of the Ministry for Innovation and Technology.

Conflicts of Interest: The authors declare no conflict of interest.

References

1. Papangelo, A.; Ciavarella, M.; Hoffmann, N. Subcritical bifurcation in a self-excited single-degree-of-freedom system with velocity weakening–strengthening friction law: Analytical results and comparison with experiments. *Nonlinear Dyn.* **2017**, *90*, 2037–2046.
2. Brockley, C.; Ko, P.L. Quasi-harmonic friction-induced vibration. *J. Tribol.* **1970**, *92*, 550–556.
3. Ibrahim, R. Friction-induced vibration, chatter, squeal, and chaos—part II: Dynamics and modeling. *Appl. Mech. Rev.* **1994**, *47*, 227–253.
4. Leine, R.; Van Campen, D.; De Kraker, A.; Van Den Steen, L. Stick-slip vibrations induced by alternate friction models. *Nonlinear Dyn.* **1998**, *16*, 41–54.
5. Hinrichs, N.; Oestreich, M.; Popp, K. On the modelling of friction oscillators. *J. Sound Vib.* **1998**, *216*, 435–459.
6. Kinkaid, N.; O'Reilly, O.M.; Papadopoulos, P. Automotive disc brake squeal. *J. Sound Vib.* **2003**, *267*, 105–166.
7. Ruina, A. Slip instability and state variable friction laws. *J. Geophys. Res. Solid Earth* **1983**, *88*, 10359–10370.
8. Dieterich, J.H. Modeling of rock friction: 1. Experimental results and constitutive equations. *J. Geophys. Res. Solid Earth* **1979**, *84*, 2161–2168.
9. Halderman, J.D.; Mitchell, C.D. *Automotive Brake Systems*; Prentice Hall: Upper Saddle River, NJ, USA, 1996.
10. Brooks, P.; Crolla, D.; Lang, A.; Schafer, D. Eigenvalue sensitivity analysis applied to disc brake squeal. In Proceeding of the Institution of Mechanical Engineering, Braking of Road Vehicles, London, UK, 23–24 March 1993.
11. Dunlap, K.B.; Riehle, M.A.; Longhouse, R.E. An investigative overview of automotive disc brake noise. *SAE Trans.* **1999**, pp. 515–522.
12. Cunefare, K.A.; Graf, A.J. *Disc Brake Rotor Squeal Suppression Using Dither Control*; Technical report, SAE Technical Paper; SAE Internationa: Warrendale, PA, USA, 2001.
13. Papangelo, A.; Ciavarella, M. Effect of normal load variation on the frictional behavior of a simple Coulomb frictional oscillator. *J. Sound Vib.* **2015**, *348*, 282–293.
14. Den Hartog, J.P. *Mechanical Vibrations*; Courier Corporation: North Chelmsford, MA, USA 1985.

15. Casalotti, A.; Arena, A.; Lacarbonara, W. Mitigation of post-flutter oscillations in suspension bridges by hysteretic tuned mass dampers. *Eng. Struct.* **2014**, *69*, 62–71.
16. Verstraelen, E.; Habib, G.; Kerschen, G.; Dimitriadis, G. Experimental passive flutter suppression using a linear tuned vibration absorber. *AIAA J.* **2017**, *55*, 1707–1722.
17. Frahm, H. Means for Damping the Rolling Motion of Ships. U.S. Patent 970,368, 13 September 1910.
18. Hamouda, M.N.H.; Pierce, G.A. Helicopter vibration suppression using simple pendulum absorbers on the rotor blade. *J. Am. Helicopter Soc.* **1984**, *29*, 19–29.
19. Sims, N.D. Vibration absorbers for chatter suppression: A new analytical tuning methodology. *J. Sound Vib.* **2007**, *301*, 592–607.
20. Habib, G.; Kerschen, G.; Stepan, G. Chatter mitigation using the nonlinear tuned vibration absorber. *Int. J. -Non-Linear Mech.* **2017**, *91*, 103–112.
21. Popp, K.; Rudolph, M. Vibration control to avoid stick-slip motion. *J. Vib. Control* **2004**, *10*, 1585–1600.
22. Chatterjee, S. On the design criteria of dynamic vibration absorbers for controlling friction-induced oscillations. *J. Vib. Control* **2008**, *14*, 397–415.
23. Niknam, A.; Farhang, K. On the passive control of friction-induced instability due to mode coupling. *J. Dyn. Syst. Meas. Control* **2019**, *141*.
24. Gantmakher, F.R. *Lectures in Analytical Mechanics*; Mir: Moscow, Russia 1970.
25. Habib, G.; Kerschen, G. Suppression of limit cycle oscillations using the nonlinear tuned vibration absorber. *Proc. R. Soc. Math. Phys. Eng. Sci.* **2015**, *471*, 20140976.
26. Kuznetsov, Y.A. *Elements of Applied Bifurcation Theory*; Springer Science & Business Media: Berlin, Germany, 2013; Volume 112.
27. Malher, A.; Touzé, C.; Doaré, O.; Habib, G.; Kerschen, G. Flutter control of a two-degrees-of-freedom airfoil using a nonlinear tuned vibration absorber. *J. Comput. Nonlinear Dyn.* **2017**, *12*.
28. Dhooge, A.; Govaerts, W.; Kuznetsov, Y.A.; Meijer, H.G.E.; Sautois, B. New features of the software MatCont for bifurcation analysis of dynamical systems. *Math. Comput. Model. Dyn. Syst.* **2008**, *14*, 147–175.

Publisher's Note: MDPI stays neutral with regard to jurisdictional claims in published maps and institutional affiliations.

lubricants

MDPI

Article

Simulation Analysis of Erosion–Corrosion Behaviors of Elbow under Gas-Solid Two-Phase Flow Conditions

Qunfeng Zeng * and Wenchuang Qi

Key Laboratory of Education Ministry for Modern Design and Rotor-Bearing System, Xi'an Jiaotong University, Xi'an 710049, China; qwch61@stu.xjtu.edu.cn

* Correspondence: qzeng@xjtu.edu.cn

check for updates

Received: 26 July 2020; Accepted: 16 September 2020; Published: 22 September 2020

Abstract: In the production and gathering process of coal gas, the complex composition of the coal gas, harsh environments, the complex medium, and high content of solid particles in slurry cause the equipment malfunctions and even failure because of erosion and corrosion. In the present study, COMSOL multi-physics finite element simulation software is used to simulate the erosion–corrosion behaviors of elbow in key chemical equipments. The electrochemical corrosion, solid particle erosion, chemical reaction, and turbulent flow are coupled together. The particle count method is proposed to clarify the erosion phenomenon. The simulation results show that particles with high turbulent intensity hit the wall of elbow directly, which forms a slanted elliptical erosion zone on the extrados surface at 40°–50°. The chemical reaction in turbulence has a difference in the concentration distribution of substances, and this phenomenon leads to different magnitudes of the corrosion current densities in the tube. Moreover, 1/6 released particles hit the extrados surface of the elbow. These findings are beneficial to understand the erosion–corrosion phenomena and design the elbow in key chemical equipment.

Keywords: elbow erosion; turbulence flow; gas-solid flow; corrosion; numerical simulation

1. Introduction

Erosion-corrosion is a comprehensive type of the failure process including the mechanical and electrochemical action [1]. The erosion–corrosion phenomena exist in many industries; however, especially in the chemical industry, serious accidents and economic losses occur easily because of the harsh working environments, corrosion, and other reasons. In 2016, an explosion caused by the leakage of the oil pipeline in the southern Gulf of Mexico's National Oil Company caused serious casualties and economic losses. Most pipelines of the coal chemical industry are made of carbon steel, which are seriously corroded by the working and environmental medium during operation [2]. For the erosion–corrosion problems, many researchers have conducted simulation studies. However, researchers performed only CFD simulation analysis, and the velocity vector of the flow filed is used to characterize the serious position of erosion [3]. In these researches, there were no comprehensive simulated erosion failure processes, such as electrochemical corrosion through current density or mass loss per unit area. Based on CFD model, the researchers have further simulated the failure process of particle erosion without electrochemical corrosion and obtained the trajectory of the particles hitting the wall [4]. Mass loss per unit area and the quantities of particles hitting the wall of elbow have not yet been obtained. Zeng et al. [5] have simulated the corrosion current density distribution of electrochemistry without the erosion and chemical reaction and acquired the galvanic current density. An erosion–corrosion simulation model involving the erosion, electrochemical corrosion, and chemical reaction under the turbulent flow condition has been rarely reported. The reason may be that the CFD analysis involves a difficult process and CFD coupling erosion or electrochemical corrosion is also difficult to develop. The corrosion module in COMSOL Multiphysics features built-in interfaces, features, and examples for modeling and analyzing these different types of corrosion. The chemical reaction of steel and fluid is simulated by chemical corrosion module, which allows us to simulate all electrochemical corrosion processes. A large amount of information can be obtained, including electrochemical reactions, potentials in electrolytes and metal structures, homogeneous chemical reactions, and unique phenomena in the corrosion process. The output of such an analysis is the localized current density, which is used to calculate the average corrosion rate of any component over a given period of time. It is also possible to observe the impact of environmental variables. Simulation analysis can assist in the identification of corrosion-related problems.

The aim is to simulate the erosion-corrosion behaviors of elbow involving the erosion of particles, electrochemical corrosions of the dissolved iron and hydrogen evolution, turbulence and chemical reactions of carbonic acid ionization simultaneously in the present paper. The complex situations in tube are simplified to gas-solid two-phase flow with the chemical reactions and electrochemical reactions, and the gas is a mixture of syngas and carbon dioxide. The Finnie's erosion model is proposed to analyze the erosion of sulfur particles eroding wall through the probability of the quantities of particles and mass loss per unit area. It is expected to obtain the turbulence characteristics, substance concentration distribution, electrochemical corrosion current density, the quantities of particles striking the wall and mass loss per unit area. The turbulent characteristics of gas in elbow and the erosion failure process of particles hitting the elbow wall are simulated under turbulence condition. The effect of turbulence on the substance concentration distribution and the substance concentration distribution on the electrochemical corrosion process of hydrogen evolution reaction and iron dissolution reaction are investigated systemically. Finally, the mechanism of different physical and chemical fields on the key parts of high temperature and pressure pipeline during operation is investigated, and the simulation model of the corrosion under the multi-field coupling actions is built. The stress, temperature, and corrosion cracks of the key parts of the pipeline between corrosion development and status are achieved, and the corrosion leakage prediction model is established and the corrosion development trend and prediction method of leakage occurrence location is predicted.

2. Simulation Model

2.1. Simplification of Chemical Equipment

Figure 1 shows a schematic diagram about the vulnerable elbow in the coal chemical plant. First, the coal-water slurry is reacted with oxygen to form the mixture gas of syngas (60%) and CO_2 (40%) at the temperature of 1350 °C. Second, the mixture gas is cooled and the temperate decreases to 250 °C in the quench chamber. Finally, the mixture gas is departed from the quench chamber and introduced into the transmission pipeline at high speed flow with solid contaminants (sulfur-containing fly ash) eroding the wall. CO_2 is dissolved in water forming the saturated carbonic acid solution that results in the electrochemical corrosion of the pipeline under the operational service or out of service conditions. The erosion-corrosion phenomena around the elbow may lead to the serious leakage under the harsh working environments.

2.2. Geometry

Three-dimensional model is built to investigate the erosion–corrosion process of elbow. Figure 2 shows a planar graph of the 90° elbow with two straight sections, four featured edges and two featured surfaces (without size scale). The lines with color are used to indicate the feature edges of the tube and analyze the process of erosion–corrosion. The diameter (D) of pipeline is 600 mm. A curvature ratio of elbow, which is the ratio of a curvature radius of the centerline to that of interior edge of the elbow, is 1.5 (dimensionless). Flow is input into the first straight section (L_1 = 3000 mm) in length at the top left corner of the figure, with a 90° elbow section and the second straight section of pipe (L_2 = 3000 mm) in length in sequence.

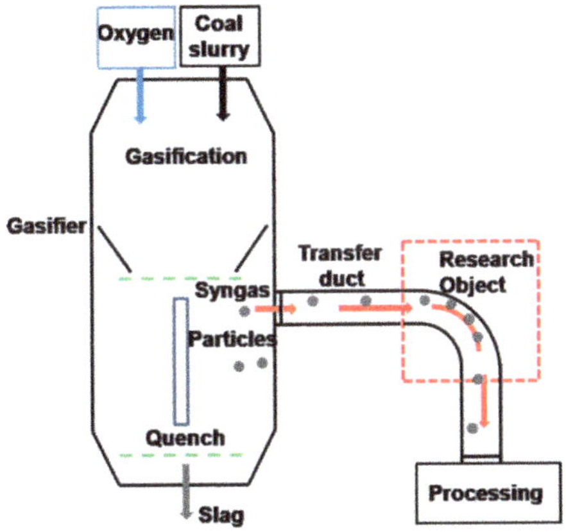

Figure 1. Schematic of chemical equipment.

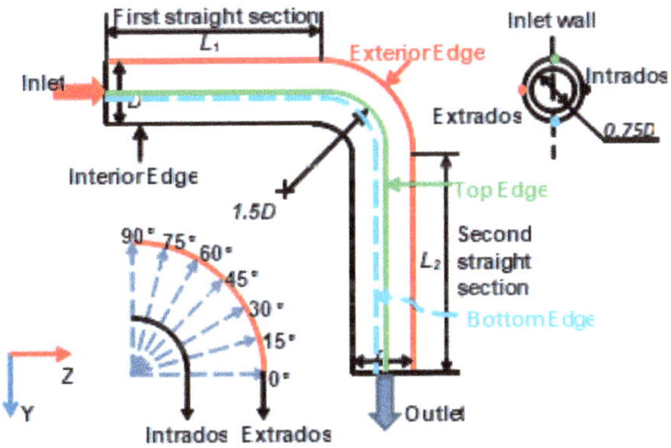

Figure 2. Schematic diagram of pipeline.

2.3. Mesh

The unstructured meshes are constructed in the computational domain. An example mesh of the pipeline is shown in Figure 3. The free triangular grid is constructed the inlet boundary (0.75 *D*). The sweep mesh is adopted because of the long narrow pipe structures and the grid processor along the pipeline to generate the structured quadrangle mesh (generate hexahedron). A boundary layer grid with 25 layers and 1.25 stretch factors is used to discretize the tube in order to ensure adequate analysis of fluid flow and particle erosion behavior near the tube wall.

Three kinds of meshes (as shown in Table 1) are constructed for the present model to test the dependence of the numerical simulation results on the mesh resolution at the same initial boundary conditions and improve the accuracy of the simulation. The variation of the max fluid velocity with the mesh resolution is examined. The results are shown in Table 1. It is found that the differences of the calculated maximum value of fluid velocity among the three meshes are very small. Thus, the first mesh method is used in this paper to reduce the calculation time.

In thin film flow, the shell interface is used to solve the Reynolds equation for flow in narrow structures and the mass and momentum balances are used to formulate with a function across the thickness of the thin structure, which indicates that the thickness does not have to be meshed. This functionality helps avoid meshing problems across the gap and thereby saves computation time.

Table 1. Mesh dependence tests.

Number	Number of Mesh Nodes	Max Fluid Velocity (m/s)
1	21,264	41.01
2	31,248	41.19
3	61,200	41.43

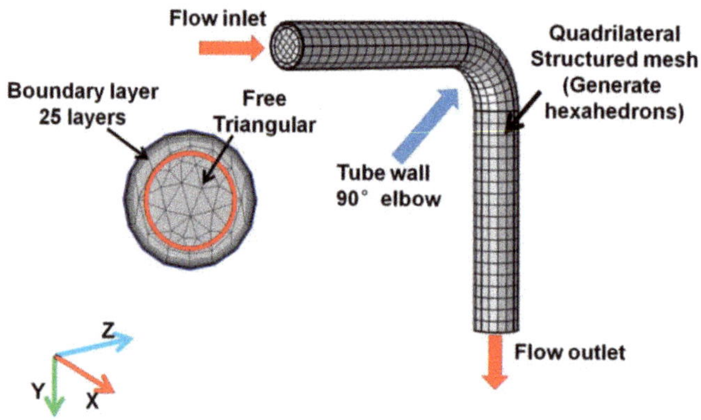

Figure 3. Mesh profile of pipeline.

2.4. The Mathematical Simulation Model

The numerical simulation model is built by a steady state solver, and then the erosion of particles is calculated by a transient solver. There are five physical field interfaces and two multi-physics interfaces in the model: turbulent flow (*spf*), particle tracking for fluid flow (*fpt*), chemistry (*chem*), secondary current distribution (*siec*), transport of diluted species (*tds*), fluid-particle interaction (*fpi*), and flow coupling (*fc*). The *chem* interface is a physical field interface in the numerical simulation software, and the *siec* interface as same.

2.4.1. Turbulence

The motion of gas is simplified to the impressible and isothermal flow according to the process conditions and gas characteristics. The k-ε model is based on the equation model and introduced an equation about the turbulent dissipation rate. The calculation is small but there is more data accumulation. It has a wide range of industrial applications with the good accuracy and good convergence. The Reynolds-Averaged Navier-Stokes (RANS) k-ε model is used to simulate and analyze the turbulence. The equations are solved on the basis of RANS equations for conservation of momentum and the continuity equation for conservation of mass in the *spf* interface. The effects of the turbulence flow are modeled through two equations—turbulent kinetic energy k and dissipation rate ε with the reliability constraints. The flow near wall is described by the wall functions. Gas is represented by a continuous phase of Euler's method and discretized using P1+P1 method. Two advanced algorithms of the streamline diffusion and crosswind diffusion are applied to the Navier-Stokes equations and turbulence equations to converge the model easily. The pseudo-time stepping algorithm is used to solve the stationary equation. The velocity of the CFL digital expression of the turbulence variable ratio parameter is 1 m/s and the length scale factor is 0.035. The specific parameters of the turbulence model are C_{e1} = 1.44, C_{e2} = 1.92, C_{μ} = 0.09, σ_k = 1, σ_c = 1.3, κ_v = 0.41, B = 5.24. These parameters are RANS k-ε model constants. The governing equations of the turbulent flow are as follows.

Incompressible flow equation in steady state:

$$\rho(U\cdot\nabla)U = \nabla\cdot[-PI + (\mu + \mu_T)(\nabla U + (\nabla U)^T)] + F \tag{1}$$

$$\rho\nabla\cdot(u) = 0 \tag{2}$$

The transport equation of k:

$$\rho(u\cdot\nabla)k = \nabla\cdot[(\mu + \frac{\mu_T}{\sigma_k})\nabla k] + P_k - \rho\varepsilon \tag{3}$$

where the production term and the turbulent viscosity are listed as following:

$$p_k = \mu_T[\nabla U : (\nabla U + (\nabla U)^T)] \quad \mu_T = \rho C_u \frac{k^2}{\varepsilon} \tag{4}$$

The transport equation of ε:

$$\rho(U\cdot\nabla)\varepsilon = \nabla\cdot[(\mu + \frac{\mu_T}{\sigma_\varepsilon})\nabla\varepsilon] + C_{\varepsilon 1}\frac{\varepsilon}{k}P_k - C_{\varepsilon 2}\rho\frac{\varepsilon^2}{k} \quad \varepsilon = ep \tag{5}$$

where μ_T, μ, ρ, P, U, I, F, k, ε, σ_k, and P_k, stand for the eddy viscosity, viscosity, density, pressure, velocity vector, unit matrix, volume force vector, turbulent kinetic energy, turbulent dissipation rate, respectively. The SI unit is used.

2.4.2. Erosion

The sulfur particles are considered as the discrete phases in the *fpt* interface during simulation. The movement of particles under the framework of Lagrange is governed by Newton's second law (6) and affected by the drag force, gravity, and brown force. The drag force (7) is generated by the speed difference between gas and particles and controlled by stokes' law (8). In addition, the body force (9) formed by the effect of accelerated or decelerated particles on the movement of gas is obtained through gas-particles interaction in the *fpi* interface, and then the gas velocity in the *spf* interface and particles velocity in the *fpt* interface are coupled by gas-particles interaction. Through sampling and analysis of the pipe ash, and in order to simplify the particle model, all sulfur particles, releasing 500 particles per 0.08 s at the boundary of inlet, are introduced into the tube by high-speed gas and there are the same physical properties of density (2360 kg/m^3), diameter (50 μm), and shape (sphere). The turbulent dispersion model of particles adopts discrete random walk, the variable time step method. The turbulent kinetic energy and turbulent dissipation rate of particles are coupled to those of gas in the *spf* interface. In addition, there is still no recognized universally applicable theoretical model due to the complexity of the erosion behavior of material. To assess the interaction between particles and pipes, Finnie erosion model is used to explain the rule of the particle erosion of plastic materials at low impact angles, thus the classical Finnie erosion model (10) is used to describe the impact of particles and the count method is used to count the quantities of particles hitting the wall considering the shape of the pipeline model. The ratio of the normal and tangential force and number multiplication factor are set to 1. The surface hardness of wall is 640 N/mm^2 and the surface mass density is 7.98 g/cm^3 respectively.

Newton's second law and drag force equation:

$$\frac{d(m_p v)}{dt} = F_D + F_g + F_{brown} \tag{6}$$

$$F_D = (\frac{1}{\tau_P})m_P(u - v) \quad (7)$$

where, F_D is Drag force, F_g is gravity, and F_{brown} is brown force. The m_p is the particle mass (SI unit: kg), τ_p is the particle velocity response time (SI unit: s), V is the velocity of particle (SI unit: m/s), and U is the fluid velocity (SI unit: m/s).

The stokes drag law for the particle response time is defined as:

$$\tau_P = \frac{\rho_P d_P^2}{18\mu} \quad (8)$$

where, μ is the fluid viscosity (SI unit: Pa·s), ρ_p is the particle density (SI unit: kg/m^3), and d_p is the particle diameter (SI unit: m).

Body force of particle-to-syngas calculation equation:

$$F_{V,j} = -\frac{1}{V_j}\sum_{i=1}^{N} n_i F_{D,i} \int \delta(r - q_i)dV \quad (9)$$

where, $F_{v,j}$ is the average volume force, a mesh element j with volume V_j, δ is the Dirac delta function, $F_{D,i}$ is the drag force exerted on the ith particle, n_i is the force multiplication factor of the ith model particle, and N is the total number of particles.

The Finnie erosion equation is listed [6].

$$\begin{aligned} V &= \frac{cMU^2}{4p(1+\frac{mr^2}{I})}[(\cos\alpha)^2] \quad \tan\alpha > \frac{P}{2} \\ V &= \frac{cMU^2}{4p(1+\frac{mr^2}{I})}\frac{2}{P}[\sin(2\alpha) - 2\frac{(\sin\alpha)^2}{P}] \quad \tan\alpha \leq \frac{P}{2} \end{aligned} \quad (10)$$

where, c (dimensionless) is the fraction of particles cutting in an idealized manner; M (SI unit: kg) is the total mass of eroding particles; U (m/s) is the magnitude of the incident particle velocity; p (Pa) is the Vickers hardness of the material; m (SI unit: kg) is the mass of an individual particle hitting the surface; r (SI unit: m) is the average particle radius; I (SI unit: kg/m^2) is the moment of inertia of an individual particle about its center of mass. For an isotropic sphere, $I = 2\,mr^2/5$; α (rad) is the angle of incidence, with $\alpha = 0$ tangent to the surface and $\alpha = \pi/2$ normal to the surface; P is a dimensionless parameter, defined as $P = K/(1 + 2mr/I)$ and K (dimensionless) is the ratio of vertical and horizontal forces action on the particle. The mass loss per unit area vs. time is chosen to evaluate the erosion rate of steel.

2.4.3. Chemical Reaction and Electrochemical Corrosion

The ionization reactions of carbonic acid are generated into the hydrogen ion and bicarbonate ions, in the *chem* interface, and the distribution of the substance concentration is affected by diffusion and convection in the *tds* interface [7]. Table 2 shows reaction equilibrium constant, positive reaction rate, and diffusion coefficient. The convection is coupled by the turbulent mixing and affected by the motion of gas depending on the mass balance Equation (11). Turbulent kinematic viscosity of substance under turbulence is governed by that of turbulent gas, and turbulent Schmidt number is 0.71.

Electrochemical corrosion is described by the reactions of dissolved iron and hydrogen evolution in the *siec* interface. The iron dissolution reaction ($Fe^{2+} + 2e^- \rightarrow Fe$) governed by anode Tafel Equation (12) occurred on the wall, and the hydrogen evolution reaction ($2H^+ + 2e^- \rightarrow H_2$) controlled by concentration-dependent kinetic Equation (13) occurred on the inner surface of the tube. The interface of metal/electrolyte is considered to be an electrode-electrolyte coupled wall to complete charge transfer and charge conservation between ions and electrons. The electrolyte conductivity is 2.5×10^{-3} S/m. The temperature in the pipeline is 250 °C. The concentration of Fe^{2+} is 1×10^{-9} mol/L. So, the concentration of H^+ is calculated by the Nernst equation, the H^+ concentration value is 1×10^{-6} mol/L. According to calculation, the initial potential of electrolyte is $E_{eq,Fe} = -0.9068$ V, and the initial potential of electrode is 0 V. The exchange current density and Tafel slope of iron dissolved reaction are 10^{-3} A/m^2 and 40 mV per decade [8], respectively. Normally, anodic oxidation of iron presents a Tafel slope of less than 60 mV. The equilibrium potential depends on Equation (14). The exchange current density of hydrogen evolution reaction is 1.1×10^{-2} A/m^2. The equilibrium potential is −0.3112 V.

Mass balance equation is:

$$\frac{\partial c_i}{\partial t} + \nabla \cdot (-D\nabla c_i) + U \cdot \nabla c_i = R_i \tag{11}$$

where, C_i is the concentration of the species (SI unit: mol/m^3), D_i is the diffusion coefficient (SI unit: m^2/s), R_i is a reaction rate expression for the species (SI unit: mol/(m^3·s)), and U is the velocity vector (SI unit: m/s).

Anode Tafel equation:

$$i_{loc} = i_0 \times 10^{\frac{\eta}{A_a}} \tag{12}$$

where, i_{loc} denotes the local charge transfer current density, i_0 denotes the exchange current density, and A_a denotes the Tafel slope.

Concentration-dependent kinetics:

$$i_{loc} = i_0[C_R \exp(\frac{\alpha_a F\eta}{RT}) - C_O \exp(\frac{-\alpha_a F\eta}{RT})] \tag{13}$$

where, i_{loc} denotes the local charge transfer current density, i_0 is the exchange current density, C_R and C_O are dimensionless expressions, describing the dependence on the reduced and oxidized species in the reaction.

$$E_{eq,Fe} = -0.44 + 2.303\frac{RT}{2F} \times \log(10^{-9}) \tag{14}$$

where, R is the gas constant, T is the absolute temperature in Kelvin, F is Faraday's constant. The Fe^{2+} concentration value of 1×10^{-9} mol/L and H^+ concentration value of 1×10^{-6} mol/L are measured and calculated by coal chemical companies.

Table 2. Constant for the model.

Constant	Source
$K_{ca} = 387.6 \times 10^{-(6.41-1.594\times10^{-3}\cdot T_f+8.52\times10^{-6}\cdot T_f^2-3.07\times10^{-5}\cdot p)}$ mol	Oddo and Tomson [9]
$k_{f,ca} = 10^{5.71+0.0526\times T_c-2.94\times10^{-4}\times T_c^2+7.91\times10^{-7}\times T_c^3}$ s^{-1}	Comprehensive chemical kinetics
$K_{bi} = 10^{-(10.61-4.97\times10^{-3}\cdot T_f+1.331\times10^{-5}\cdot T_f^2-2.624\times10^{-5}\cdot P)}$ mol	Oddo and Tomson
$k_{f,bi} = 10^9$ s^{-1}	Nordsveen [10]
$D_{H2CO3} = 2.00 \times 10^{-9}$ m^2/s	Kvarekval [11]
$D_{HCO3^-} = 1.105 \times 10^{-9}$ m^2/s	Newman [12]
$D_{H^+} = 9.312 \times 10^{-9}$ m^2/s	Newman
$D_{CO3} = 0.92 \times 10^{-9}$ m^2/s	Kvarekval

Note: In the table, T_f is temperature in degrees Fahrenheit, T is absolute temperature in Kelvin, T_c is temperature in degrees Celsius, P is the absolute pressure, D is the diffusion coefficient, equilibrium (K) and forward (k_f) reaction rate coefficients.

3. Results

3.1. Turbulence Characteristics of Gas

3.1.1. Pressure, Friction Speed Characteristics, and Turbulence Intensity

Figure 4a,b shows the pressure distribution contour of the pipeline. Although the pressure in the entire pipe is set to a constant (6.5 MPa), the pressure on the extrados surface (6.51 MPa) is higher than that on the intrados surface (6.49 MPa). Figure 4c,d shows the distribution of the friction speed along the wall. The friction speed on the elbow surface is usually higher than that on the wall of straight sections. The friction speed on the intrados surface of elbow is higher than that on the extrados surface. The friction speed on the intrados surface of the first half of elbow is higher than that of the second half, however, the distribution of friction speed on the extrados surface is contrasting. Figure 4e,f shows the turbulence intensity distribution of gas. The turbulence intensity is strong and complex at the elbow section. In addition, the turbulent intensity coil is formed into a concave shape in the second straight section, and the coil points to the intrados surface.

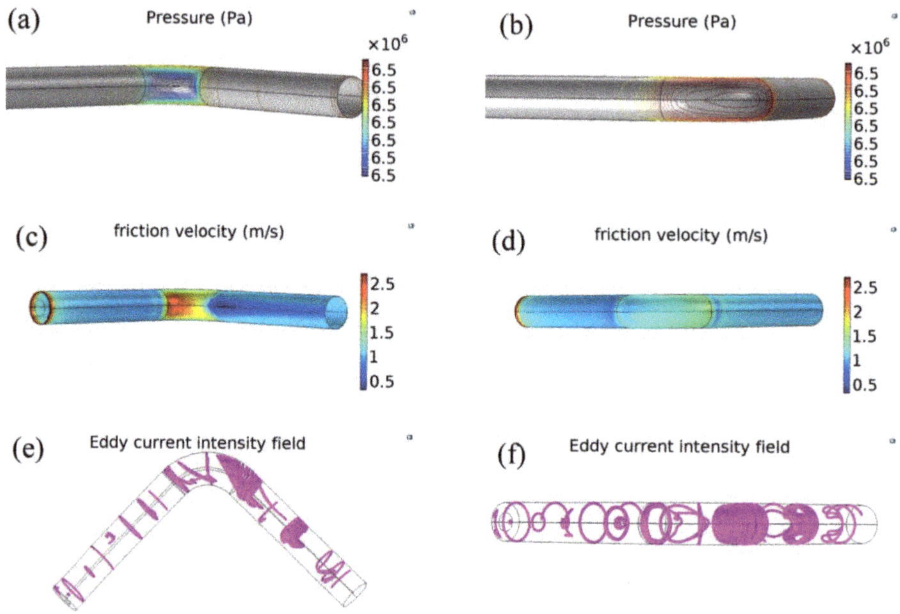

Figure 4. Turbulence characteristic. (**a**,**b**) Pressure contour at different views; (**c**,**d**) friction velocity distribution on the wall at different views; (**e**,**f**) turbulent strength at different views.

3.1.2. Velocity Streamline in Pipeline

Figure 5 shows the distribution of streamlines in pipeline, where the color of the streamline indicates the velocity magnitude. From Figure 5a, the maximum flow velocity (41.01 m/s) and the minimum flow velocity (1.609 m/s) are concentrated in the elbow section. From Figure 5b, the location of maximum flow velocity and minimum flow velocity is close to the intrados surface. The velocity of gas increases and then decreases near the intrados surface, however, the velocity of the gas decreases and then increases near the extrados surface. From Figure 5b,c, the streamlines gradually were concentrated and many flow lines were terminated in the second straight section. Furthermore, most of streamlines are concentrated near the intrados surface and formed a ε-shape distribution. From Figure 5d, the streamlines near the extrados in elbow are constrained by the wall and diverged into the second straight section, and the middle streamlines has a relatively high velocity.

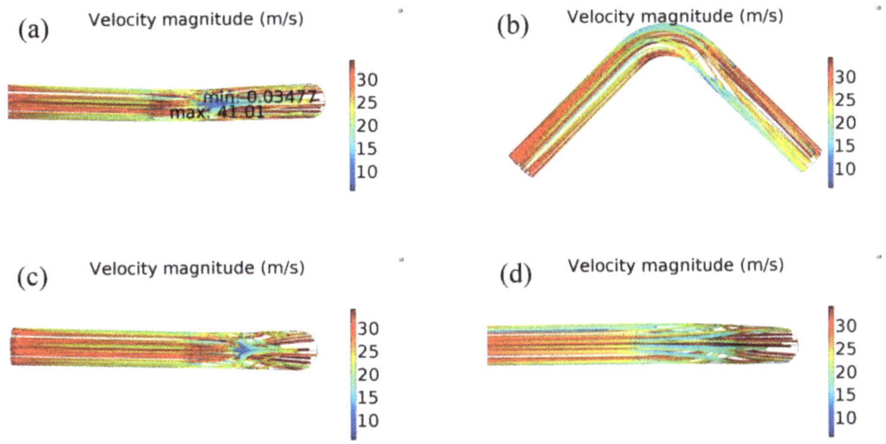

Figure 5. A three-dimensional diagram of fluid velocity distribution at different views (speed in color).

3.1.3. Velocity along the Four Featured Edges

Figure 6 shows a velocity diagram that indicates the variation in velocity along the four edges of pipeline. Near the inlet, the velocity along four edges drops from 30 m/s to about 13 m/s, and then a small reduction in speed occurs in the first straight section. Near the elbow section, the flow velocity on the interior edge was subject to small fluctuations. Although the velocity magnitude of interior edge in the elbow section has increased dramatically (up to 28 m/s) and then undergoes a drastic reduction (low as 1 m/s), the flow velocity on other three edges is increased. In the second straight section, the speeds on the top and bottom edges decrease slightly and stay within a certain value, while the speed on the exterior edge still maintains an increase, and then reduced and kept consistent. On the interior edge, speed has been gradually increased to reach steady state.

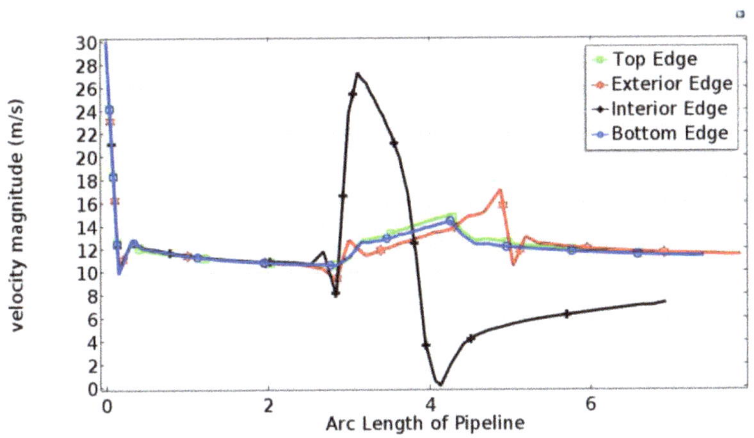

Figure 6. Line graph of velocity magnitude along four edges of pipeline.

3.2. Electrochemical Corrosion Behavior

3.2.1. Species Concentration Distribution Characteristic

Figure 7 shows the concentration distribution of chemical substances in the pipe. The concentration of substances is accumulated at the entrance of the elbow, especially near the intrados surface.

Carbonic acid concentration reaches 443.84 mol/m^3. Hydrogen ion concentration is maintained at a constant value of 9.70 mol/m^3. The highest concentration of carbonate ions reaches 9.71 mol/m^3. The maximum concentration of bicarbonate ion is 5.55 mol/m^3. However, near the extrados surface, the accumulation of substance concentration is relatively small. All simulations were performed by the experimental parameters in Table 1.

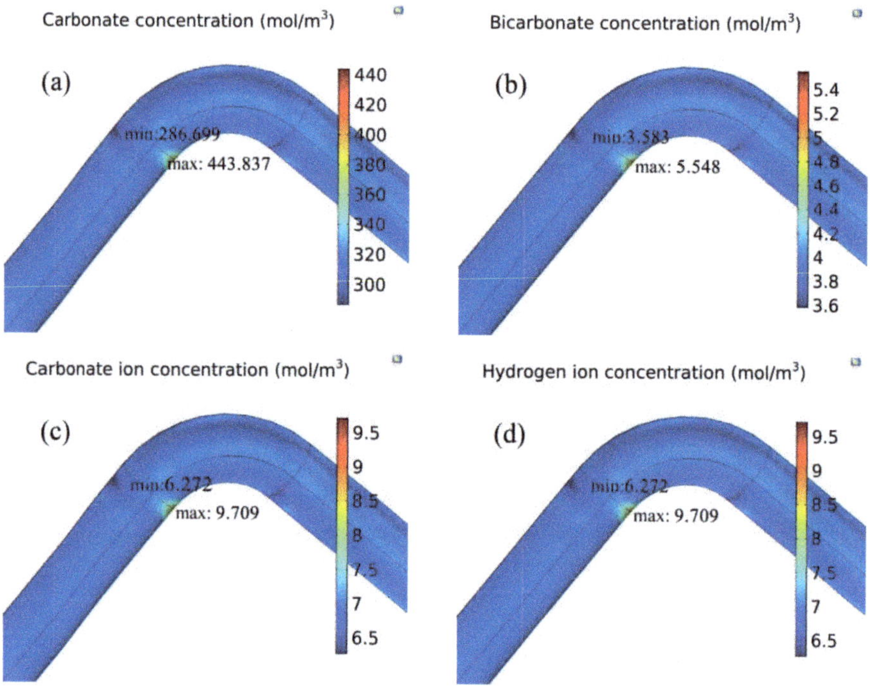

Figure 7. A three-dimensional diagram of material concentration distribution. (**a**) Carbonic acid concentration distribution; (**b**) bicarbonate concentration distribution; (**c**) carbonate ion concentration distribution; (**d**) hydrogen ion concentration distribution.

3.2.2. Current Density Characteristic

Figure 8 shows the variations in current density of four feature edges along the wall from the inlet to the outlet of the pipe. Figure 8a shows the total current density of the interface; Figure 8b shows the anode current density and Figure 8c shows the cathode current density.

Total Interface Current Density of Four Edges

From Figure 8a, the current density of four edges hardly has a difference in the first straight section. However, in the elbow section, the current density on the interior and exterior edges has a sharp fluctuation respectively. The current density on the exterior edge has a relatively positive fluctuation in the first half of elbow, and the current density on the interior edge has a negative and a positive fluctuation respectively. In the second straight section, the current density of four edges has a small fluctuation around 0 A/m^2.

Anode Current Density of the Four Edges

From Figure 8b, the anode current density of the top and bottom edges is maintained about 4.4×10^{-4} A/m^2 throughout the pipeline. However, the current density on the interior and exterior edges of the elbow section has a sharp fluctuation, especially on the interior edge. The anode current on the interior edge increased sharply to about 6.1×10^{-4} A/m^2 in the first half of the elbow, then was maintained at about 4.4×10^{-4} A/m^2, and finally decreased to 4.1×10^{-4} A/m^2 in the second half of the elbow. The anode current density on the exterior edge decreased to 3.9×10^{-4} A/m^2 in the first half of elbow, and then maintained a stable value at around 4.4×10^{-4} A/m^2.

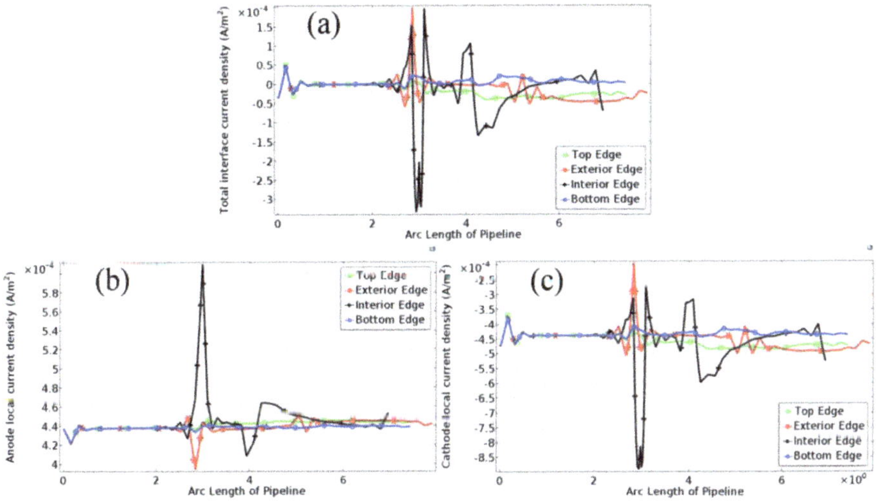

Figure 8. The current density distribution of four featured edges on the tube. (**a**) Total interface current density distribution; (**b**) anode current density distribution; (**c**) cathode current density distribution.

Cathode Current Density of the Four Edges

From Figure 8c, the cathode current density of four edges in the first straight section maintains at about -4.4×10^{-4} A/m^2. However, in the elbow section, the current density on the interior and exterior edges has a sharp fluctuation respectively. The current density on the interior edge in the first half of elbow has a relatively negative fluctuation (about -9.0×10^{-4} A/m^2) and then in the second half of the elbow has a relatively positive fluctuation (about -2.7×10^{-4} A/m^2). The current density on the exterior edge in the first half of elbow has a positive fluctuation (about -2.4×10^{-4} A/m^2). In the second straight section, the current density on four edges has a small fluctuation around -4.5×10^{-4} A/m^2.

Current Density Distribution on the Wall

Figure 9 shows the current density distribution caused by electrochemical reactions under turbulence throughout the pipeline. From Figure 9a, the serious part of the total interface current density is mainly concentrated at the intersection of the elbow and the straight sections. From Figure 9b, the most serious metal loss caused by the electrochemical corrosion is concentrated at the junction of the elbow and the first straight section. From Figure 9c, the serious part of the cathode local current density is similar to the part of the total interface current density.

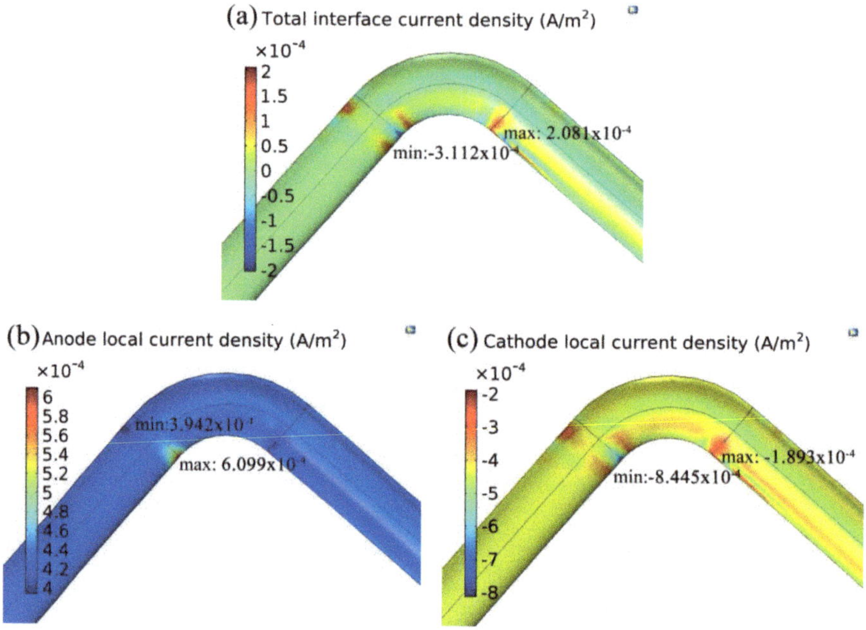

Figure 9. Current density 3D map on the wall. (**a**) Total interface current density; (**b**) anode local current density; (**c**) cathode local current density.

3.3. Erosion Behavior

3.3.1. Particle Trajectory

Figure 10 shows the trajectory of the particle motion in the entire pipeline and Figure 11 shows the trajectory of particle hitting the wall of the elbow, where the color of the particles indicates the velocity. Form Figure 10a,b, particles in the first straight section basically maintained the same velocity, except for particles with the minimum speed near the extrados surface. In addition, at the junctions of elbow and the first straight section, the velocity of the solid particles was changed, and particles were significantly moved toward the extrados surface. From Figure 10c, the particles in the first half of the elbow hit the extrados surface along an approximately straight line, and particles near the extrados surface were gradually pushed toward intrados surface. As shown in Figure 10c–e, many particles hit the extrados surface of elbow. As shown in Figure 11, the particles hit the elbow forming a curved distribution and bouncing off the wall. A small quantity of particles lagged near the intrados when particles are at the junction of the elbow and the second straight section. In Figure 10f–h, the motion of the particles in the second straight section gradually became more dispersive, and the particles formed the cloud cluster.

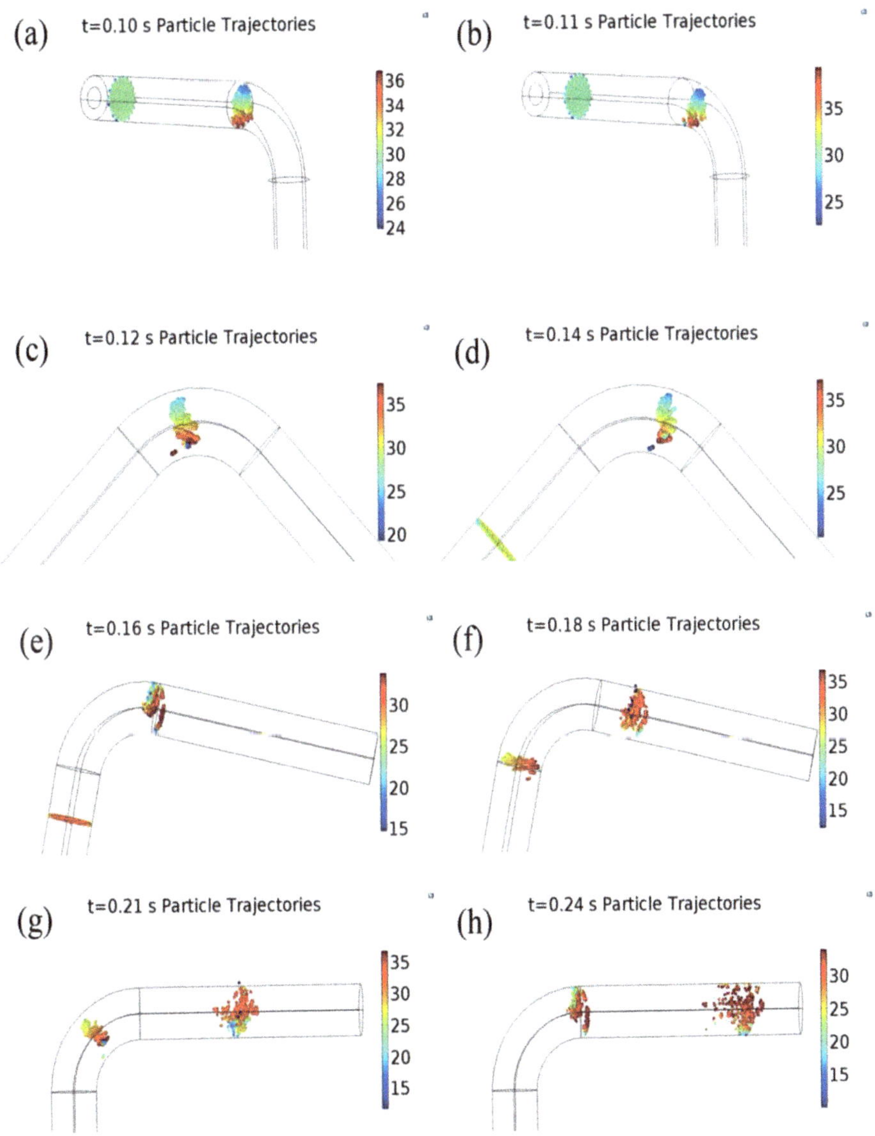

Figure 10. Three-dimensional diagram of particle trajectories in the pipeline (the color indicates the speed). (**a**) t = 0.10 s particle trajectories; (**b**) t = 0.11 s particle trajectories; (**c**) t = 0.12 s particle trajectories; (**d**) t = 0.14 s particle trajectories; (**e**) t = 0.16 s particle trajectories; (**f**) t = 0.18 s particle trajectories; (**g**) t = 0.21 s particle trajectories; (**h**) t = 0.24 s particle trajectories.

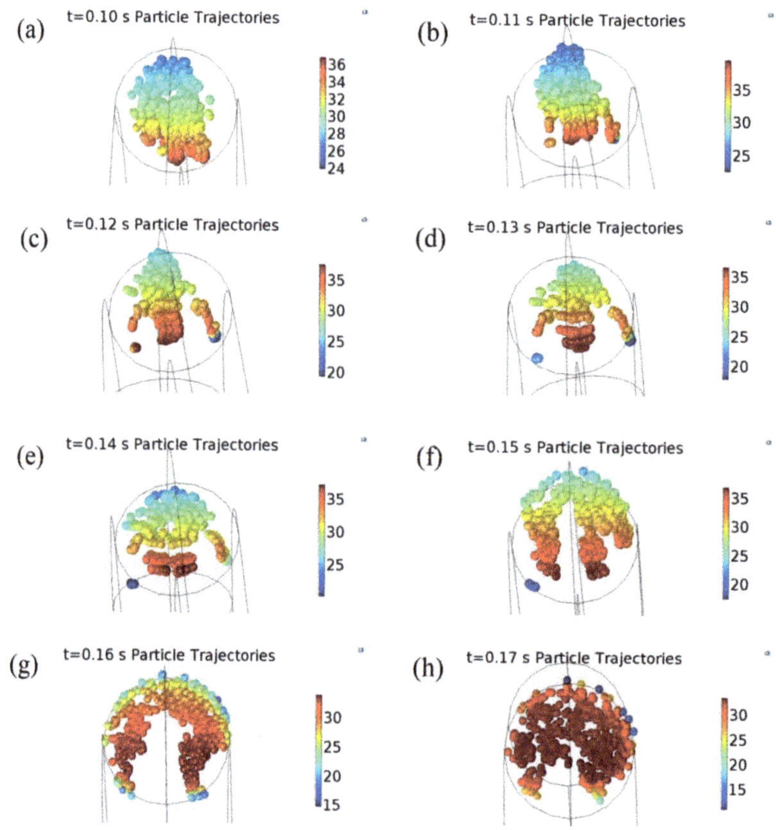

Figure 11. The trajectory of particles hitting the wall (the color indicates the speed). (**a**) t = 0.10 s particle trajectories; (**b**) t = 0.11 s particle trajectories; (**c**) t = 0.12 s particle trajectories; (**d**) t = 0.13 s particle trajectories; (**e**) t = 0.14 s particle trajectories; (**f**) t = 0.15 s particle trajectories; (**g**) t = 0.16 s particle trajectories; (**h**) t = 0.17 s particle trajectories.

3.3.2. The Quantities of Particles Striking the Wall and Mass Loss Per Unit Area

Figure 12 shows the variation of quantities of particles hitting the wall. Figure 13 reveals that the variation of mass loss per unit area caused by the erosion. During the simulation, the count method is used to study the particles collision wall. The results are shown in Table 3. Numbers of solid particle collisions is around 0.2×10^4 of the maximum value at the straight section, and numbers of solid particle collisions is around 1.255×10^4 of the maximum value at the elbow. It can be concluded that approximately 16.7% (1/6) of the total released particles hit the extrados of the elbow and caused a serious loss of quality. As time goes on, the quantity of particles striking the wall gradually increases, accompanied by an increase in the loss of metal quality. In addition, it can be found that the mass loss per unit area mainly occurs between 40° and 50°, with the most serious region between 43° and 48°, and gradually was spread to the surrounding area, forming oblique elliptical erosion area. The simulation results considering the presence of acidic substances can be verified from other papers [13].

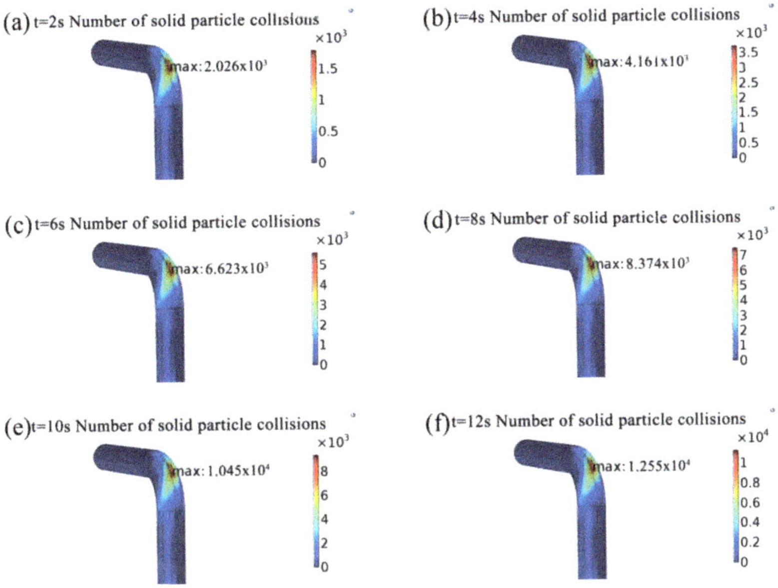

Figure 12. The number of solid particles hitting the wall. (**a**) t = 2 s number of solid particle collisions; (**b**) t = 4 s number of solid particle collisions; (**c**) t = 6 s number of solid particle collisions; (**d**) t = 8 s number of solid particle collisions; (**e**) t = 10 s number of solid particle collisions; (**f**) t = 12 s number of solid particle collisions.

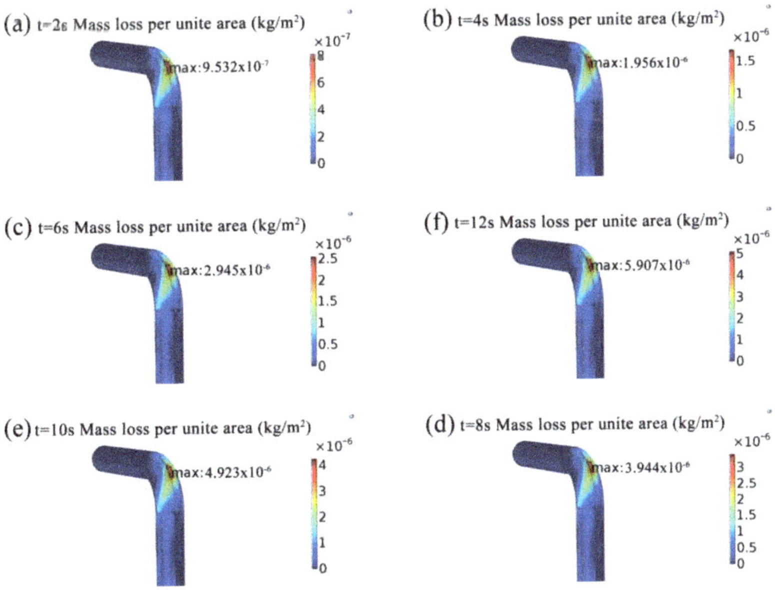

Figure 13. The evolution of mass loss per unit area (kg/m^2). (**a**) t = 2 s mass loss per unit area; (**b**) t = 4 s mass loss per unit area; (**c**) t = 6 s mass loss per unit area; (**d**) t = 8 s mass loss per unit area; (**e**) t = 10 s mass loss per unit area; (**f**) t = 12 s mass loss per unit area.

Table 3. Particle collision data over time.

Time	The Total Number of Particles in the Pipeline	The Number of Particles Hitting the Extrados	Statistical Probability of Collision	Maximum of Mass Loss per Unite Area on the Extrados (kg/m^2)
2 s	12,500	2026	16.2%	9.53×10^{-7}
4 s	25,000	4161	16.6%	1.96×10^{-6}
6 s	37,500	6263	16.7%	2.95×10^{-6}
8 s	50,000	8374	16.75%	3.94×10^{-6}
10 s	62,500	10,450	16.72%	4.92×10^{-6}
12 s	75,000	12,550	16.73%	5.91×10^{-6}

4. Discussions

Pipeline is an important part of equipments, which is widely used in petrochemical, aerospace, and other industrial applications, and is extremely important in safe production. There are unavoidable erosion and corrosion during applications, which maybe lead to leakage of the pipeline because of harsh working environments [14–18]. The main reasons of leakage are erosion, electrochemical corrosion, turbulent and chemistry, and their interactions. Corrosion is one of the most damaging mechanisms in many engineering materials and structures. Erosion of the pipelines may result in the failure of the piping system, which can be extremely dangerous and expensive. Erosion may enhance corrosion and corrosion may enhance the erosion rate through preferential dissolution and this is the so-called synergistic effect. It is well-known that, because of the synergistic effect, the possibilities of leakage are generally much higher than the sum of pure electrochemical corrosion and pure mechanical erosion. Moreover, elbows are the weak parts of gathering and transferring pipelines. In studies of erosion–corrosion there are no models available which attempt to combine the effects of particle erosion, electrochemical corrosion, chemistry, and fluid flow [19]. Nowadays, because of a strong demand for the analysis of erosion–corrosion problems, this study focuses on the following aspects: numerical study on the fluid flow, erosion, and corrosion along the axial direction of the pipeline for the real field cases and the related phenomenon of erosion–corrosion expecting to provide detailed and reasonable analysis of the failure incurred by the erosion–corrosion.

4.1. The Correlation between Streamline and Turbulence

Although the problems caused by the synergistic effect of erosion–corrosion are serious, the erosion–corrosion mechanism of the elbow, as influenced by the velocity and pressure, is still not thoroughly understood because of its complexity. Numerical simulations are often used in erosion–corrosion research [20]. The velocity difference on the four featured edges near the inlet is due to high wall lift force. The slow decline of speed in the first straight section is attributed to the fact that the characteristics of the boundary layer flow field and the turbulent core area is quite different. Boundary layer has blocking effect on the fluid motion and the gradient of each flow parameter is very large leading to the formation of lift and affects the velocity of the wall.

However, as a result of the variation of pressure and turbulent intensity, the speed on four edges has different fluctuations throughout the elbow. The pressure variation, as shown in Figure 4a,b, is ascribed to the cause that constraint in geometry results in the gases to accumulate in the elbow section, and the accumulative effect exerts a relatively high pressure on the extrados and moderately relieves the pressure on the intrados. The intense turbulent intensity are due to the truth that low viscosity of gas causes vertical flow of gas: the fluid here produces an axial velocity and a radial velocity perpendicular to the tangential velocity of the main gas flow under centrifugal force. Because the very various velocity of the fluid causes different centrifugal forces in the elbow, so here the fluid exhibits complex three-dimensional motion characteristics. The research indicates that the places of pipeline including connection affect the erosion–corrosion behaviors [21]. The change in geometry causes a sharp change in the direction of flow with a large variation in velocity and friction action or retardation on wall surface. As shown in Figure 4c,d, in the elbow section, the gases were forced to change direction with high friction velocity, which affects the fluid motion and causes low speed near the interior surface. The reduction in the amount of streamline may be due to the intense turbulence in the elbow that causes the gas movement to concentrate in some streamlines in the second straight section.

In the second straight section, the effect of turbulence intensity still affects the motion of gas, but the movement will gradually become stable along with the distance increased and the turbulence intensity dissipated.

4.2. The Correlation of the Electrochemical Corrosion with Material Concentration Distribution

The model includes the electrolyte domain and the electrode surface. The concentration distribution of substances produced by carbonic acid ionization reactions is mainly concentrated at the junction of the elbow and the first straight section where turbulence intensity dramatically increased. As shown in Figure 4e,f and Figure 7, the intense turbulent intensity at the entrance of the elbow is a barrier to mass transfer, and the substance is difficult to transfer to intense turbulence intensity region and cause substance to be accumulated. Furthermore, turbulent intensity is very complex and intense at the elbow and it causes the substance to pass quickly and the remaining influence continues to the outlet boundary. The flow field characteristics in the boundary layer play a role in retarding the transport of the material concentration, thus causing the material to accumulate near the wall surface.

The variation in the electrochemical corrosion current density is a concentration-dependent process in a local region, and electrochemical reaction depends on the charge transfer reaction between the ions in the electrolyte and the electrons on the wall. The charge is conserved throughout the process. Convection and diffusion affect the concentration distribution of substance. The electrochemical corrosion current density is related to substance concentration distribution. Therefore, high current density of anode on the interior edge and low current density of anode on the exterior edge of the elbow is related to the concentration distribution of material.

4.3. Erosion in the Turbulence

It is well-known that erosion has an important role in the total erosion–corrosion rates [22]. Severe erosion occurred at elbow is caused by the impact of solid particles driven by high-speed gas. At elbow, the direction and magnitude of gas was changed sharply because of the low viscosity and geometric constraint, and gas has a relatively small effect on the movement of particles. When the inertia of the solid particles is relatively large, the solid particles can pass through the streamline and hit the wall surface almost in a near linear path. For the straight sections, the random collision and erosion of the tube wall are caused by particles because of the influence of the pulsation of the flow field, but this effect is small. Therefore, the most severe erosion occurred at the elbow. In addition, the movement of the particles near the wall surface is lagged because of the boundary layer blocking effect. The geometric constraint and the effect of inertia of particles cause the particle group to form a curve near the extrados surface because of the boundary layer close to the tube wall and different turbulence characteristics.

There is an elliptically eroded area on the extrados surface of the elbow due to the presence of gravity, Brownian force, drag forces, and the turbulent intensity variations. In the first straight section, particles are gradually affected by the nature force; the trajectories of the particles gradually offset the line and gradually moved toward the extrados surface and the gravity direction. At the elbow, the particles impacted the wall and reduced their kinetic energy, which resulted in the subsequent particle group to be pushed and accumulated. In addition, the drastic changes in the flow direction, and the increase in turbulence intensity, caused the particles in the boundary layer to hit the wall surface and to be scratched along the wall surface, resulting in an elliptical erosion region from the bottom area to the top area. In addition, the formation of the cloud cluster in the second straight section is because the particles have different velocity directions and magnitudes at the entrance, which is affected by the wall constraint and high turbulence intensity at the elbow.

5. Conclusions

Erosion-corrosion behaviors of elbow are simulated and the model containing erosion, electrochemistry corrosion, turbulence, and chemistry is built to describe the complicated failure phenomena. The conclusions are summarized as follows:

1. The serious erosion of elbow occurs between 40° and 50°, and gradually expanded into the surrounding area forming a slanted erosion region under the force, boundary layer's blocking action and wall rebound. The particles at high speed hit the wall of the tube, especially in the elbow section with high turbulent intensity, which caused the serious erosion area on the extrados of elbow.
2. The particles count method is proposed to describe erosion and provide a probability prediction of the elbow lifetime. About 16.7% particles collided the extrados surface during erosion.
3. The corrosion current density of iron is concentrated in the junction of the straight section and elbow and the intersection of the straight section and the intrados surface of elbow. The strong turbulent intensity in elbow and the boundary layer affects the substance concentration of chemical reactions accumulated at the junction of the straight section and elbow.
4. The pressure on the extrados is higher than that on the intrados. However, the friction speed on the intrados is higher than that on the extrados due to the cumulative effect of gas in pipe. Low viscosity of gas and the geometric constraints are attributed to cause difference in the velocity magnitude and particles hitting the wall in straight line.

Author Contributions: Q.Z. conceptualized the study and administered the project; Q.Z. and W.Q. performed the simulations; Q.Z. and W.Q. analyzed the data and wrote the manuscript; Q.Z. revised this article. All authors have read and agreed to the published version of the manuscript.

Funding: The present work is financially supported by the key project of Shaanxi province Science and Technology Department (2017ZDXM-GY-115) and National Natural Science Foundation of China (51675409), Natural Science Basic Research Plan in Shaanxi Province of China (2019JM-274) and the Open Project Program of Beijing Key Laboratory of Pipeline Critical Technology and Equipment for Deepwater Oil & Gas Development (Grant No. BIPT2018001).

Acknowledgments: The author would like to thank the Shaanxi Special Equipment Research Institute and coal gasification companies in northern Shaanxi for the valuable data. At the same time, thanks to all the funds provided.

Conflicts of Interest: The authors declare no conflict of interest.

References

1. Rajahram, S.; Harvey, T.; Wood, R. Erosion–corrosion resistance of engineering materials in various test conditions. *Wear* **2009**, *267*, 244–254. [CrossRef]
2. De Waard, C.; Lotz, U.; Milliams, D. Predictive model for CO_2 corrosion engineering in wet natural gas pipelines. *Corrosion* **1991**, *47*, 976–985. [CrossRef]
3. Liang, G.; Peng, X.; Xu, L.; Cheng, Y.F. Erosion-corrosion of carbon steel pipes in oil sands slurry studied by weight-loss testing and CFD simulation. *J. Mater. Eng. Perform.* **2013**, *22*, 3043–3048. [CrossRef]
4. Liu, H.; Zhou, Z.; Liu, M. A probability model of predicting the sand erosion profile in elbows for gas flow. *Wear* **2015**, *342*, 377–390. [CrossRef]
5. Zeng, L.; Shuang, S.; Guo, X.P.; Zhang, G.A. Erosion-corrosion of stainless steel at different locations of a 90 elbow. *Corros. Sci.* **2016**, *111*, 72–83. [CrossRef]
6. Finnie, I. Some observations on the erosion of ductile metals. *Wear* **1972**, *19*, 81–90. [CrossRef]
7. Patterson, C.S.; Slocum, G.H.; Busey, R.H.; Mesmer, R.E. Carbonate equilibria in hydrothermal systems: First ionization of carbonic acid in NaCl media to 300 °C. *Geochim. Cosmochim. Acta* **1982**, *46*, 1653–1663. [CrossRef]
8. Nesic, S.; Postlethwaite, J.; Olsen, S. An electrochemical model for prediction of corrosion of mild steel in aqueous carbon dioxide solutions. *Corrosion* **1996**, *52*, 280–294. [CrossRef]

9. Oddo, J.; Tomson, M. Simplified calculation of $CaCO_3$ saturation at high temperatures and pressures in brine solutions. *J. Pet. Technol.* **1982**, *34*, 1583–1590. [CrossRef]
10. Nordsveen, M.; Nešic, S.; Nyborg, R.; Stangeland, A. A mechanistic model for carbon dioxide corrosion of mild steel in the presence of protective iron carbonate films—Part 1: Theory and verification. *Corrosion* **2003**, *59*, 443–456.
11. Kvarekval, J. A kinetic model for calculating concentration profiles and fluxes of CO_2-related species across the nernst diffusion layer. In *Corrosion 97, March 9–14, 1997*; NACE International: New Orleans, LA, USA, 1997.
12. Newman, J.; Thomas-Alyea, K. *Electrochemical Systems*; John Wiley & Sons: Hoboken, NJ, USA, 2012.
13. Zhang, E.; Zeng, D.; Zhu, H.; Li, S.; Chen, D.; Li, J.; Ding, Y.; Tian, G. Numerical simulation for erosion effects of three-phase flow containing sulfur particles on elbows in high sour gas fields. *Petroleum* **2018**, *4*, 158–167. [CrossRef]
14. Xu, Y.; Tan, M. Probing the initiation and propagation processes of flow accelerated corrosion and erosion corrosion under simulated turbulent flow conditions. *Corros. Sci.* **2019**, *151*, 163–174. [CrossRef]
15. Zhao, J.; Xiong, D.; Gu, Y.; Zeng, Q.; Tian, B. A comparative study on the corrosion behaviors of X100 steel in simulated oilfield brines under the static and dynamic conditions. *J. Pet. Sci. Eng.* **2019**, *173*, 1109–1120. [CrossRef]
16. Islam, M.A.; Farhat, Z.N.; Ahmed, E.M.; Alfantazi, A.M. Erosion enhanced corrosion and corrosion enhanced erosion of API X-70 pipeline steel. *Wear* **2013**, *302*, 1592–1601. [CrossRef]
17. Wood, R.J.K.; Wharton, J.A.; Speyer, A.J.; Tan, K.S. Investigation of erosion–corrosion processes using electrochemical noise measurements. *Tribol. Int.* **2002**, *35*, 631–641. [CrossRef]
18. Zeng, L.; Chen, G.; Chen, H. Comparative study on flow-accelerated corrosion and erosion–corrosion at a 90 carbon steel bend. *Materials* **2020**, *13*, 1780. [CrossRef] [PubMed]
19. Stack, M.; Abdelrahman, S.; Jana, B. A new methodology for modelling erosion–corrosion regimes on real surfaces: Gliding down the galvanic series for a range of metal-corrosion systems. *Wear* **2010**, *268*, 533–542. [CrossRef]
20. Liu, J.; BaKeDaShi, W.; Li, Z.; Xu, Y.; Ji, W.; Zhang, C.; Cui, G.; Zhang, R. Effect of flow velocity on erosion–corrosion of 90-degree horizontal elbow. *Wear* **2017**, *376*, 516–525. [CrossRef]
21. Jianwen, Z.; Aiguo, J.; Yanan, X.; Jianyun, H. Numerical investigation on multiphase erosion-corrosion problem of steel of apparatus at a well outlet in natural gas production. *J. Fluids Eng.* **2018**, *140*. [CrossRef]
22. Okhovat, A.; Heris, S.Z.; Asgarkhani, M.H.; Fard, K.M. Modeling and simulation of erosion–corrosion in disturbed two-phase flow through fluid transport pipelines. *Arab. J. Sci. Eng.* **2014**, *39*, 1497–1505. [CrossRef]

lubricants

MDPI

Review

Review on Friction and Wear Test Rigs: An Overview on the State of the Art in Tyre Tread Friction Evaluation

Andrea Genovese *, Gennaro Antonio D'Angelo, Aleksandr Sakhnevych and Flavio Farroni

Department of Industrial Engineering, University of Naples Federico II, Via Claudio, 21-80125 Naples, Italy; gennar.dangelo@gmail.com (G.A.D.); ale.sak@unina.it (A.S.); flavio.farroni@unina.it (F.F.)

* Correspondence: andrea.genovese2@unina.it

Received: 22 July 2020; Accepted: 11 September 2020; Published: 16 September 2020

Abstract: The future evolution of autonomous mobility and road transportation will require substantial improvements in tyre adherence optimization. As new technologies being deployed in tyre manufacturing reduce total vehicle energy consumption, the contribution of tyre friction for safety and performance enhancement continues to increase. For this reason, the tyre's grip is starting to drive the focus of many tyre developments nowadays. This is because the tread compound attitude to maximize the interaction forces with the ground is the result of a mix of effects, involving polymer viscoelastic characteristics, road roughness profiles and the conditions under which each tyre works during its lifespan. In such a context, mainly concerning the automotive market, the testing, analysis and objectivation of the friction arising at the tread interface is performed by means of specific test benches called friction testers. This paper reviews the state of the art in such devices' development and use, with a global overview of the measurement methodologies and with a classification based on the working and specimen motion principle. Most tyre friction testers allow one to manage the relative sliding speed and the contact pressure between the specimen and the counter-surface, while just some of them are able to let the user vary the testing temperature. Few devices can really take into account the road real roughness, carrying out outdoor measurements, useful because they involve actual contact phenomena, but very complex to control outside the laboratory environment.

Keywords: friction testers; tribometers; viscoelastic materials; rubber friction; tyre

1. Introduction

Tribology is the science and technology concerning the interaction of solid surfaces in relative motion. The word tribology derives from the Greek word "tribos" that means rubbing. The topics covered by this word are various and include the study of lubricants, lubrication, friction, wear and bearings [1]. Leonardo Da Vinci was one of the first to carry out and to report studies in the field of tribology, at the end of the XVth century; he had not only performed experimental studies concerning friction but he had also developed diverse ingenious schemes for the measurement of friction. His work remained unpublished until the twentieth century when Dowson [2] presented his monumental study regarding the history of tribology. It is fascinating to note how the Da Vinci's studies on friction still remain scientifically significant today [3]. An example of Leonardo's sketches published by Dowson is shown in Figure 1.

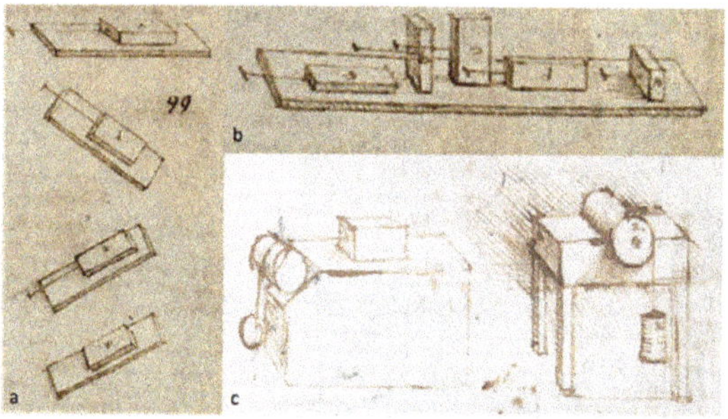

Figure 1. Sketches from Leonardo's notebooks: (**a**,**b**) from Codex Atlanticus (Biblioteca Ambrosiana, Milan, Italy; CA folio 532r c. 1506-8), and (**c**) from Codex Arundel (British Library, London, UK; Arundel folio 41r c. 1500-05) [3].

Although this research topic has been studied for centuries, new analytical, numerical, and experimental methods have continued to evolve and be developed due to the intrinsic difficulty of the examination of materials' tribological properties. Indeed, they do not only depend on the type of material and the relative properties, but also on the geometry, surface conditions and topography. In addition, the measurements are also affected by several working conditions such as the pressure distribution within the contact interface, relative speed, sliding distance, temperature and relative humidity [4–6]. Consequently, extensive experimental studies adopting the most effective and robust methodologies become absolutely necessary for a deep understanding of tribological phenomena. To overcome the measurement and testing difficulties, a large amount of testing devices has been developed in the last centuries. These devices, variously called tribometers, tribotesters or friction testers (FTs), are widely used to study the friction phenomena of completely different materials, with a particular interest towards viscoelastic ones due to their advantageous characteristics, intrinsically variable in different application working ranges, and therefore particularly suitable for vibration and noise isolation or impact and impulsive shock absorption.

Starting from a particular interest in the study and in the characterization of the viscoelastic materials, the review aims to illustrate and to discuss the experimental devices designed and developed for the study of the rubber friction, being of crucial aspect in completely different contexts, e.g.,: shoe soles, O-ring sealing, conveyor belt and automotive applications, where the vehicle dynamics, the design of the apposite control systems, and the performance- and safety-focus are largely affected by what happens at the tyre/road interface in terms of friction generation mechanisms [7–10].

The review focuses on the devices that aim to study the friction between tyres and the road. It is worth highlighting that there is no limitation on the type of rubber that can be used for the friction and wear tests to be carried out with such kinds of devices, but the most commonly exploited testing conditions simulate the typical working contact conditions in the field of tyres, which are not necessarily reflected in other fields or standards. As for the latter point, it should be noted that there are some differences between the two main standards, the International Standards Organization (ISO) and the International American Standard for Testing and Materials (ASTM). For example, although in ASTM there are several indications concerning the proper selection of a method to measure the friction properties of a generic material, there are no any particular specifications for the determination of the rubber friction properties [11]. In ASTM G115-10, the "Standard Guide to Measure and Report Friction Coefficients" is also included in the ASTM Friction Test Standards so that users can choose which method may be most suitable for a particular application [12]. The methods for the determination of rubber friction are instead described in the ISO 15113 standards. This international standard refers to a linear movement and, unlike the previous ones, does not describe in details the test apparatus, but rather only provides a guide on the experimental arrangement, procedures and on the parameters to be taken into consideration to perform a robust measurement pipeline. Furthermore, the ISO standard gives indications about the normal loads and speeds to be used within testing, additionally providing procedures for the preliminary preparation of sliding surfaces under analysis [13].

In the majority of theoretical studies on the frictional properties of materials, friction is represented using the friction model developed by Amontons and Coulomb, who claim that the frictional force is proportional to the normal force or load. However, as demonstrated initially by Bowden and Tabor [14,15] and by other authors later, Coulomb friction models are not fully reliable in case of viscoelastic materials like rubber [16,17]. Indeed, the main characteristics of the rubber friction, nowadays widely accepted and experienced, are dependent on normal force and sliding speed, temperature and a real contact area with highly-non-linear relations. These parameters play a key role in the study of the frictional behaviour of viscoelastic materials and can be used as a classification criterion to distinguish the devices and the studies to be conducted on rubbers and those to be performed in the tyre field. For instance, in a great amount of studies on rubber friction, tests are conducted at very low experimental speeds (lower than 1 mm/s) [18], so that the temperature effect in the contact area can be neglected [19] while tests conducted within the typical tyre working conditions should be generally performed at speeds in the order of meters per second.

Another crucial objective of a modern tribometer employable for tyre studies should consist in giving the possibility to work with specific samples of countersurface, i.e., an asphalt sample, since there could be a significant amount of reasons affecting the pavement characteristics: mix designs, plant operations, existing pavement conditions, or operations of the paver. It is quite obvious that it becomes really difficult to obtain the desired pavement characteristics in laboratory since even particular paving operations or peculiar mixture transformation during the compaction process phase may deeply modify the countersurface characteristics and therefore the viscoelastic behaviour of the tyre rubbers during the tyre-road interaction. For this reason, depending on a testing facility employed, it is always recommended to make sure to work with the pavement surface as similar as possible to the real one, extracting samples from a real tarmac surface for indoor testing or allowing to perform the analyses directly on track for outdoor testing with the aim to reproduce completely real test conditions. In case the test rig is employable for outdoor testing, it should be preferred since it allows one to not alter the geometries under study.

As already mentioned, the knowledge of the tyre friction behaviour and of the parameters affecting the phenomenon is an important topic both for academicians and industrial researchers, involving crucial aspects such as safety, performance, durability and environmental concerns [5]. To this end, a series of test benches have been developed by universities, research institutions and tyre makers. The laboratory tests, in addition to being less expensive compared to outdoor tests, offer the possibility to carry out measurements in almost completely controlled environmental conditions, allowing one to vary sliding speed, normal load, temperature, and other parameters in wide ranges. Another important advantage of testing tyre tread block elements in the laboratory lays in the fact that the investigations can be made at a very early stage of the tyre development, when an eventual change within the compound composition is still relatively cheap and can be easily performed, optimizing as a consequence the tyre tread geometry of sipes or other design features. The goal of the test benches is to carry out tests, not only hardly reproducible outdoors, but also representative and transferable to large-scale tyre testing. A previous investigation of these devices was conducted by Moldenhauer [20] in 2010 as part of his Ph.D. thesis. The work done by Moldenhauer has been deeply analysed, and, in the authors' opinion, further enrichment is required in order to provide a more complete analysis of the devices developed in recent years. Furthermore, Moldenhauer's study is limited to the description from the constructive point of view, while this review aims to analyse also the experimental outputs obtained, with the aid of published research references.

The paper is organized as follows: firstly, an insight on the theoretical aspects of the rubber friction is reported to provide the reader with a panorama of the main approaches and to point out the most critical aspects; then an overview of the experimental devices developed in the last fifteen years is illustrated. Such an overview cannot be complete, since there is a large number of devices used in universities' research departments or in the tyre industry, not always accessible, but it still represents a quite complete report of the current techniques and methodologies, with particular reference to dry contact conditions. Tribometers can be classify based on the contact mechanism/geometry (area, line or point contact) of the tested material with the counter-face [21]; the type of motion of the moving part [22] (linear, rotary/rolling, reciprocating, or a combination); the element motion-actuated. Referring to the type of motion, Sections 3–5 are respectively dedicated to Rolling FTs, Linear FTs and Other Types, which includes the devices that are not attributable to the other two categories.

2. Theoretical Description of Rubber Friction

The friction properties of elastomers such as rubber have been extensively studied for decades [23–26]. The tribological properties of rubber depend on many parameters, e.g., surface roughness, speed, normal load, lubrication, temperature and material properties. A fundamental study in this field was conducted by Grosch in his pioneering work [25], where different types of rubber were experimentally analysed on diverse hard surfaces, mentioning two distinct processes taking part at the generation of the friction phenomena: the adhesion, akin to a molecular relaxation process, and a deformation process in which energy is lost due to the cyclic stress of the rubber due to the surface roughness [19], also known as the hysteretic component of friction. The friction coefficient evaluations obtained from experimental studies or mathematical models are usually plotted on a graph as a function of the sliding speed, this kind of representation is also called a "master curve". This curve can be also parametrized for other variables such as load and temperature. The master curve of rubber on a rigid rough surface exhibits, in general, two distinct peaks. The first, being attributed to the adhesive component of friction, occurs in general at low sliding velocity, whereas the second one, referring to the hysteretic contribution occurs at higher sliding velocity [27]. A typical example of a master curve is represented in Figure 2, where the friction coefficient for a styrene butadiene rubber (SBR) rubber sliding on three different surfaces is reported.

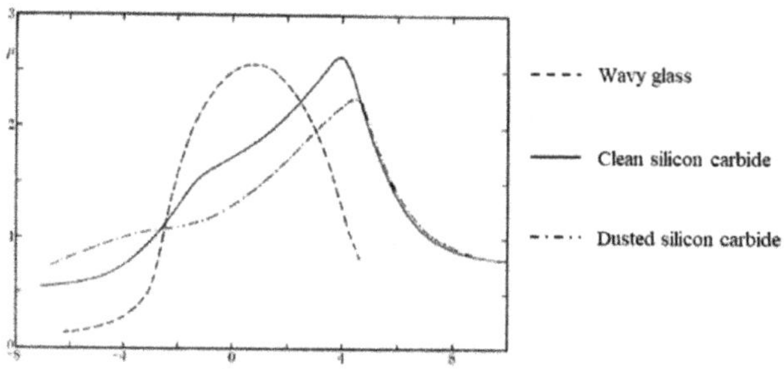

Figure 2. Master curve of the friction coefficient for SBR rubber sliding on three different surfaces [25].

Subsequently, other researchers such as Greenwood and Williamson (GW) [28] introduced a concept involving the contribution of the true contact area on friction. In particular, the formulation of the GW theory is based on Archard's [29] previous idea of multi-asperity contact. They approximated the roughness of the surface as an ensemble of spheres having the same radius randomly distributed over the mean plane to take into account of the surface statistics.

Over the years, some other studies such as the ones by Bush at al. [30,31], Heinrich, Kluppel and others [32–34] proposed further contributions based on the original GW theory. Multi-asperity contact models, based on Greenwood and Williamson theory, represent one of the two most used approaches to account for the true contact area on the friction mechanisms. Another widely adopted approach has been developed by Persson [35], whose theory, in contrast to the GW models, removes the assumption that the true contact area is smaller than the nominal contact area [36], considering the extreme case of full contact conditions between a rigid rough surface and an initially flat elastic half-space. Such theory takes partial contact into account requiring that, in case of adhesionless contacts the stress probability distribution vanishes when the local normal surface stress is fading. It also assumes that the power spectral density (PSD) of the deformed elastic surface is the same as the rough surface below [37]. The theory provides formulas, needing as inputs only the PSD surface and the elastic properties of the contacting bodies. A recent study by Carbone and Bottiglione [38] compared the two different approaches, stating that Persson's rubber contact and friction theory, and the subsequent theories based on it, are more accurate. Although theories on rubber friction have evolved over the years, they still present diverse limitations associated with the adhesive and the viscoelastic component of friction [39]. Under this light, appears evident the centrality of an experimental approach and how important the setup of the different testing fixtures is.

3. Rolling Friction Testers

This category of devices makes use of the principles of double disc tribometers, which consist of two discs rotating against each other. This kind of tribometers are also known with different names such as: ring on ring, roll on roll, rolling sliding apparatus, etc. In the field of tyres, these tests equipment reproduce, on a small scale, the design of the "tyre on drum" test machine.

Liu et al. [40,41] have developed a high-speed rolling test rig aiming to simulate the impact and release mechanisms of tread block. As shown in Figure 3, the device consists of a small wheel with a rubber belt coating, which drives a big wheel with a steel surface. The small wheel is mounted directly on a solid base and is driven by an electric motor, while the big wheel is mounted on a solid base by a moving rocking arm; lastly another couple of stiff arms are used to lock the wheel shaft position, to obtain the desired value of interference between the two wheels and to apply a compression to the tread block sample.

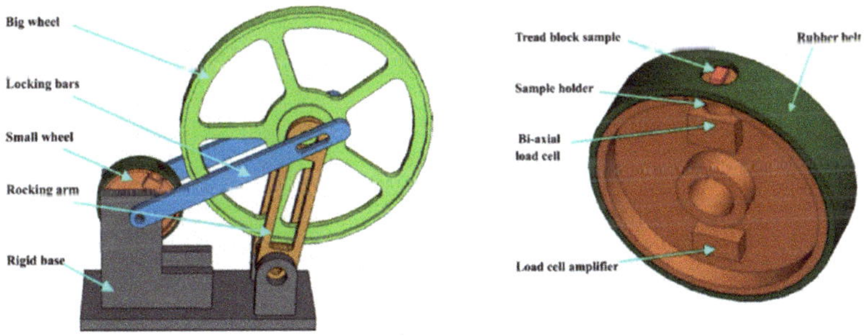

(a) (b)

Figure 3. Rolling test rig (**a**), main component of driving wheel (**b**) [41].

The main elements of the driving wheel are:

- an electric motor, driven by an inverter (DF51, Moeller,)
- a rubber belt, attached around the outside surface, with a hole to allow installation of the tread block sample
- a biaxial load cell embedded inside and interposed between the tread block and the sample holder
- a load cell amplifier
- a thermocouple placed in contact of the sample, used to measure operating temperature
- a slip ring unit

The main technical specifications are summarized in the Table 1.

Table 1. High-speed rolling test rig technical specification.

	Min	Max	
Speed	-	600	(rpm)
Wheel interference	0.1	0.3	mm
Temperature	-	Amb	(°C)
Tread block sample	16 × 10 × 6.35		(mm^3)

With the aim of the investigation regarding the contact forces between the tread block and the road, the authors conducted several tests, varying both the speed (at three speed levels of 150, 300, 600 rpm), and the interferences in the 0.1–0.3 mm range.

With the high-speed rolling test rig, it is possible to simulate the impact and release mechanisms of a tread block. The main drawback of this device is related to the use of the asphalt surface, for two reasons: firstly, it is difficult to make a curved asphalt sample; secondly, when the big wheel is covered by a rough surface, the tread-block sample comes in contact with the wheel only partially.

A different approach for rolling FT is proposed by Lundberg et al. [42], involving a new experimental device that allows detailed studies of the rolling contact force between a tread block and a relative substrate. The device, called compact internal drum (CID), aims to simulate a realistic impact and release mechanism for the tread block-substrate contact and enables force measurements.

Figure 4 shows the core of the device, consisting of two wheels: a solid metal wheel (1) with a cut-out window to accommodate a force link (2) interposed between the tread block sample and the inner face of the wheel. The tread sample is obtained by cutting a tread strip from a truck tyre re-treading material, like the one shown in Figure 5. The solid wheel rolls on the inner surface of the second wheel (drum) (3), that can be covered by an interchangeable rough surface.

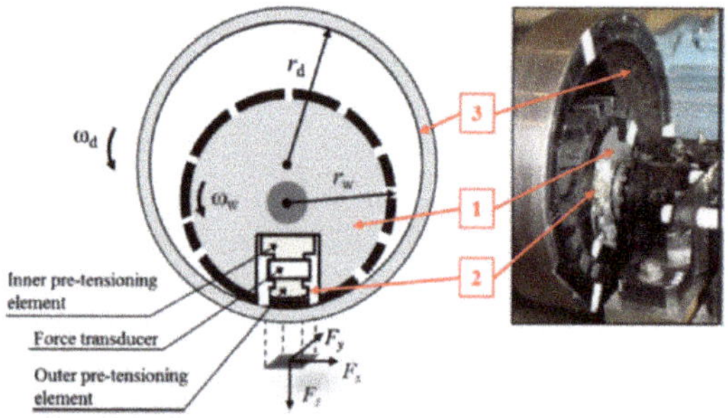

Figure 4. The core of the test rig [42].

(a) (b) (c)

Figure 5. Truck tyre retreading material (**a**), tread strip (**b**) tread strip main dimension (**c**) [42].

In Figure 6, the test rig is depicted in all its components, the highlighted (numbered) components complete the device description. The solid wheel is driven by an electric motor (4), controlled by a driver, which transmits the movement directly to the shaft of the solid wheel, through three belts, and subsequently from this to the drum through the friction forces established between the rubber mounted on the periphery of the solid wheel and on the inner surface of the drum. The drum shaft is equipped with an automotive-derived disc brake (5), used to deaccelerate the drum; furthermore, it is possible to vary the brake pressure by means of a manually adjustable hydraulic system (9). An optical sensor (7) is used to measure rotation speeds, while a resistive temperature sensor is introduced into a tread block adjacent to the sample tread block, allowing to estimate the real temperature without damaging the reference sample. The data from the sensors are transmitted from the rotating part to the non-rotating data acquisition system via an HBM SK12 slip ring (8). It is possible to prescribe the desired compression load between the tread block sample and the rough surface of the drum, by shifting vertically the shaft of the drum. The normal load between the sample of the tread block and the rough surface of the drum is quantified by measuring the bending deformation induced in the strain gauges mounted on the drum shaft. The test rig is mounted on a concrete block (6). The main technical specifications of the rig are listed in Table 2.

Figure 6. Layout of the test rig [42].

Table 2. Compact internal drum technical specification.

	Min	Max	
Speed	0	17	(m/s)
Normal force	0	300	(N)
Temperature	-	Amb	(°C)
Tread block sample	30 × 34 × 20		(mm^3)

By adopting this kind of test rig the contact forces between tread block and roads under different conditions can be investigated. Indeed, the test rig layout allows to vary: the velocity, the normal load and the braking pressure, with or without driving torque. In. [42], Lundberg et al. showed the experimental results obtained varying the main parameters involved.

The main advantage in the use of the CID lies in the possibility to operate in a wide range of speeds; furthermore, if compared with the device analysed in the previous section, it offers the possibility to measure the forces generated in the tyre/road contact both in free-rolling conditions or in sliding conditions, simply varying the braking force. Again, a drawback of this layout is related to its geometry that imposes the use of curve asphalt samples which are particularly difficult to produce.

4. Linear Friction Testers

The first linear friction tester (LFT) was developed by Leonardo da Vinci at the end of the fifteenth century and the concept of his device is still valid and adopted in recent studies [43,44]. Indeed, in their recent study [45], Tolpekina and Persson analysed adhesion and friction in tyre tread compound, conducting an experimental investigation trough a LFT like the Leonardo da Vinci type. The schematically representation of the tester is available in Figure 7.

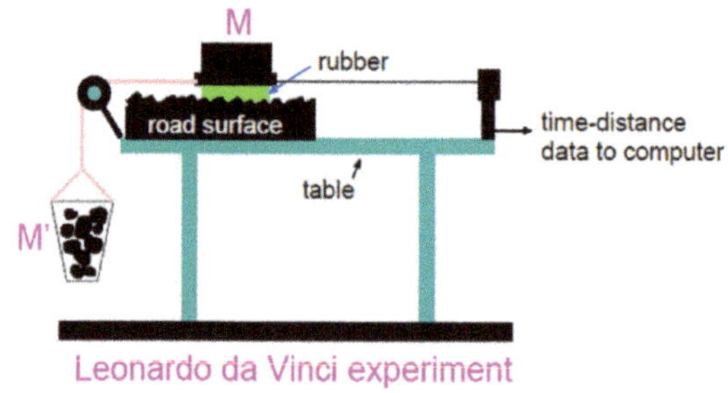

Figure 7. Schematic representation of the Leonardo da Vinci set-up [45].

The layout of the device is as simple as smart: it consists of a table on which a substrate is fixed and on which a rubber sample can slide. The latter can be loaded by means of a calibrated weight and moved by adding a weight in a container connected to its free end. Nowadays this type of test rigs is widely used, perhaps is the most used device in the study of friction and wear in tyre tread block.

In the modern era, Grosch's tribometer is considered one of main examples of this category of testers [25]. The device, shown in Figure 8, consists of a test sample pad of about 2.5 cm^2 and 0.5 mm thick, attached to a sample holder and pressed against the test surface by a calibrated load. To measure the frictional force a U-spring dynamometer is connected to the holder by means of two steel wires. The test surface is driven by a motor coupled to a multi-ratio gear box, that provides sliding speeds ranging from 10^{-6} to 3 cm/s. The sample and the track surface are enclosed in a temperature and humidity-controlled chamber.

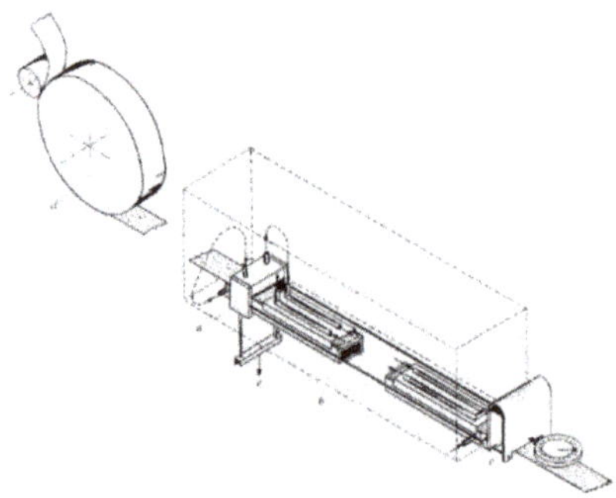

Figure 8. Schematic representation of Grosch's tribometer. (**a**) Specimen holder, (**b**) narrow tunnel for preheating the surfaces, (**c**) dynamometer, (**d**) self-tightening drive, coupled to gear box and motor; normal load applied at (**e**) [25].

Over the years, researchers have adopted various layouts to improve the friction coefficient measurements, evolving from the principles developed by Grosch's works. Given the wide range of existing LFT devices for convenience, in the following, they are distinguished in two categories:

- LFT in which the moving part is the rough surface

- LFT in which the moving part is the rubber sample/specimen

4.1. LFT Surface in Motion

The layout of LFT in which the moving part is the rough surface is a setup adopted by numerous researchers Lorenz et al. [46,47] carried out a rubber friction study for a tread rubber sample sliding on an asphalt road specimen taking advantages from this kind of setup.

More in detail, the friction measurements have been carried out using an in-house developed test rig, schematically shown in Figure 9, composed by a lower steel sledge where the rough surface sample is clamped. The sledge is moved using a voice coil actuator, capable to generate a constant force. To control the actuator and its speed, the position of the sledge is measured using an analog magneto-strictive linear position encoder. On the upper side of the device there is an aluminium plate which on one side (the lower surface) allows to attach the rubber sample; on the other (upper surface), encapsulates a heating system; in this way it is possible to ensure a homogeneous distribution of the heat, ensured by inserting the device in a temperature-controlled chamber where the temperature can be controlled. The nominal normal load can be varied through the application of calibrated weights on the upper plate, allowing the contact force between tread sample and rough surface to be changed. The friction force is measured using a bi-axial load cell mounted in line with the sample holder. The main technical specifications are shown in Table 3.

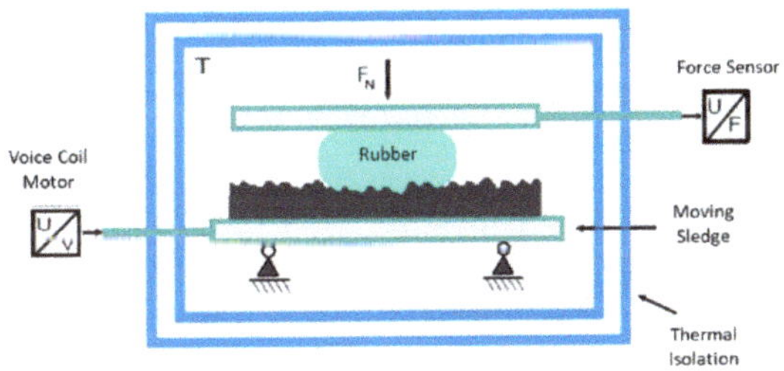

Figure 9. Schematic representation of LFT surface in motion devices.

Table 3. Technical specification of the LFT adopted by Lorenz et al.

	Min	Max	
Speed	5×10^{-6}	1×10^{-3}	(m/s)
Normal force	-	26	(N)
Temperature	−10	+120	(°C)
Tread block sample	$20 \times 20 \times 5$		(mm^3)

With the LFT, the authors aimed to investigate contact forces between tread blocks and road surfaces. The tests were conducted under different sliding conditions, varying both the velocity in the range $10^{-6} < v < 10^{-3}$ m/s and temperature at three different levels, −8 °C, 20 °C, 48 °C, respectively. The rubber test sample is cut out from a tread compound used on summer tyres for passenger car. The measured friction forces are then shifted according to the Williams, Landel and Ferry (WLF) equation [48].

It is worth noting that this device allows measurements only in a small range of velocities and loads, therefore it is not suitable to simulate the real conditions of tyre/road contact. A similar LFT layout was adopted by Lang and Kluppel [49], designed for the experimental investigation of the load and temperature dependences upon the dry friction behaviour of racing tyre tread compound in contact with rough granite. The device employed for the study was constructed and developed at IMKT, University of Hannover. The scheme of the device is shown in Figure 10.

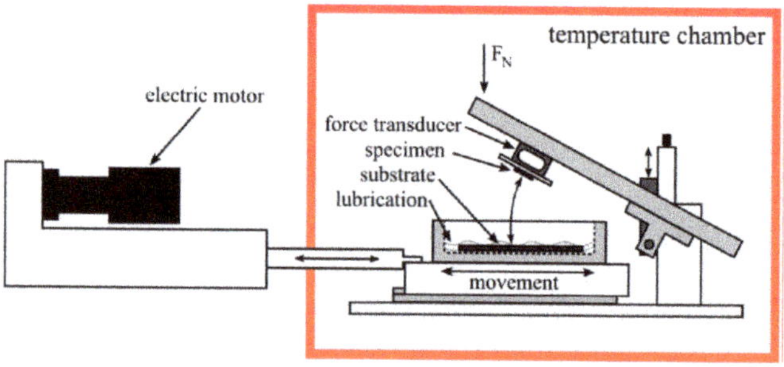

Figure 10. Schematic representation of the friction tester used for measurements [49].

The test rig is composed by the following main elements: An electric motor; an arm on which a rubber specimen is housed; a force transducer, interposed between the rubber sample and the arm; a tank, moved by the motor through an actuator, to host a road specimen; a granite road specimen and a temperature chamber. The nominal normal force acting on the tread sample can be varied. Through the application of different values of the load, on the upper side of the arm, it is possible to get a different value of the contact pressure between tread sample and rough surface. The main technical specifications of the test rig are shown in Table 4.

Table 4. IMKT, University of Hannover LFT technical specification.

	Min	Max	
Speed	1×10^{-4}	1	(m/s)
Normal force	-	7	(bar)
Temperature	2	100	(°C)
Tread block sample	$20 \times 20 \times 8.5$		(mm^3)

The tests, aimed to investigate the friction coefficient, were conducted under different sliding conditions, varying both the velocity and the temperature aspects. Measurements in different velocity conditions from 0.1 mm/s to 300 mm/s have been carried out changing the load between 1 bar and 7 bar at six different temperatures (at 2 °C, 10 °C, 20 °C, 40 °C, 70 °C and 100 °C).

The rubber test sample was cut from a tread compound of a racing tyre and moved in sliding contact on two different granite surfaces: coarse and fine. In [49] is reported the friction coefficient carried out form the experimental activity, both for the coarse and for the fine granite, at different load levels in the range 1–7 bar and for six different reference temperature.

Compared to the other FTs presented in the review, the device adopted by Lang and Kluppel seems to be one of the most complete since it allows one to operate in both dry and wet conditions; to investigate the effects of the utilization of interface lubricants and the use of real asphalt and tread samples. Moreover, a wide range of applicable loads and the temperature control environment allows measurements in almost all tyre-road contact operating conditions of passenger or racing automotive applications.

The last device belonging to this category is reported in O'Neil et al.'s [50] studies, where the authors performed an experimental investigation to predict tyre behaviour on different road surfaces. To perform friction measurements, the authors have used a LFT. The device, constructed at the University of Surrey (Guildford, UK), has a layout very similar to the previous one, reported in its schematic representation and in the real arrangement in Figures 11–13.

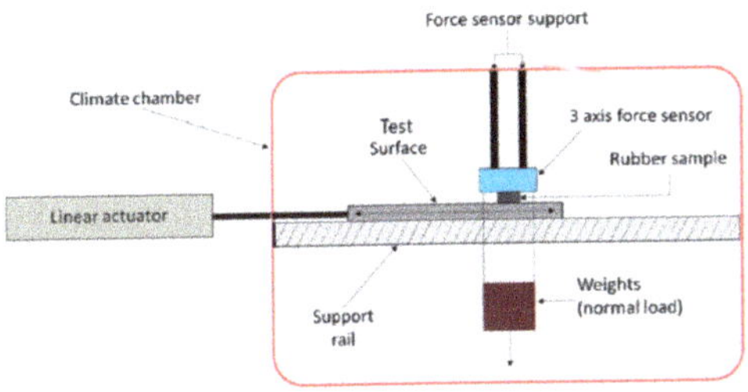

Figure 11. Schematic representation of the friction tester at University of Surrey.

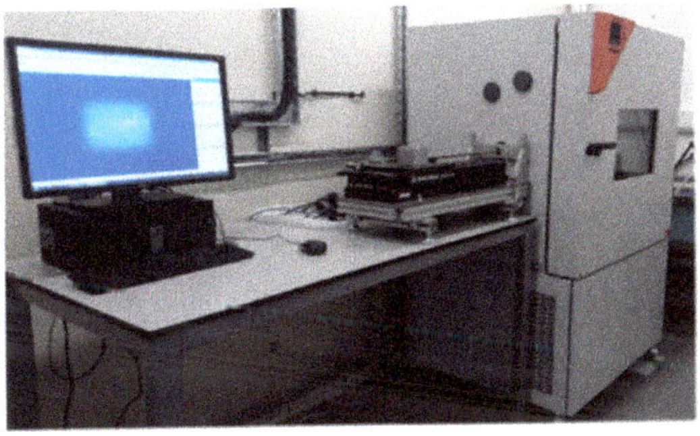

Figure 12. View of the LFT at University of Surrey [51].

Figure 13. Internal test chamber view [51].

A direct-drive linear motor, located outside the climate chamber, moves a sledge where the test surface is clamped. The sledge is embedded in the climate chamber and is supported by a linear rail system. The rubber sample is placed above the test surface by a rigid frame, between the sample holder and the frame is interposed a three-axis force sensor. The tread sample is pressed against the test surface through the application of calibrated weights on the upper side of the sample holder; in this way it is possible to vary the contact pressure between tread sample and rough surface. The main technical specifications are summarized in the Table 5.

Table 5. Technical specification of the LFT at University of Surrey.

	Min	Max	
Speed	1×10^{-6}	0.05	(m/s)
Temperature	−40	+180	(°C)
Tread block sample	35×40		(mm^2)

In [50], tests were conducted on a rubber sample cut from a passenger car tyre and driven on sandpaper. Measurements have been carried out for different velocities from 0.03 mm/s to 10 mm/s at sixteen different temperatures ranging from −40 °C to +50 °C. The main technical difference between the above test rig and the one at IMKT (University of Hannover), regards only the maximum sliding speed reachable (about 0.05 m/s).

4.2. LFT Rubber Sample in Motion

In the current section a second family of LFT is analysed, characterized by the fact that the moving part is the tread block sample. A first example of this kind of layout is reported in the studies conducted by Lahayne et al. [52,53], in which the results of the friction coefficient measurements are carried out using a LFT developed at the Institute for Mechanics of Materials and Structures at Vienna University of Technology [54]. The schematic representation and the core of the test bench are shown in Figure 14.

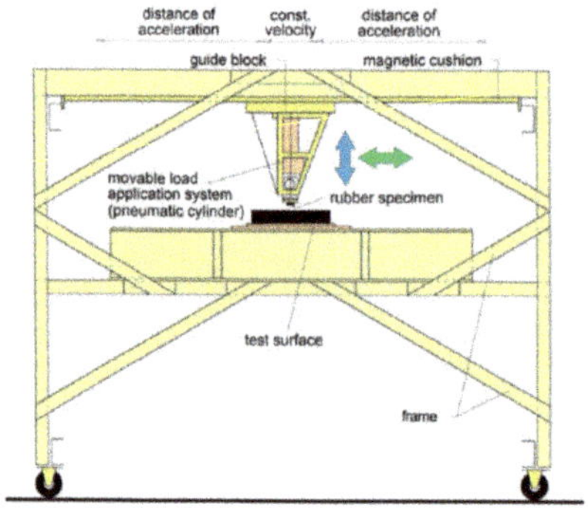

Figure 14. Schematic representation of LFT at Vienna University (adapted from [55]).

The LFT consists of a rigid frame that contains both interchangeable test surfaces and a linear motion unit. The linear unit holds and pulls a sledge that incorporates both the tread block specimen and the test equipment. This sledge contains: a pneumatic system used to generate the normal load, a sample holder and a piezoelectric force sensor which measures the contact force between tread sample and rough surface. The frame was specially designed to hold a high-speed camera and a pyrometer to acquire the deformation and the temperature of the sample during the tests. The device is located in a climate chamber.

The main technical specifications are summarized in Table 6.

Table 6. Technical specification of the LFT at Vienna University.

	Min	Max	
Speed	1×10^{-5}	1.1	(m/s)
Normal force	-	30	(bar)
Temperature	−30	+40	(°C)
Sliding distance	-	300	(mm)
Tread block sample	$80 \times 20 \times 10$		(mm^3)

The tests were performed by the authors under different sliding conditions, varying velocity, temperature, rubber materials and rough surfaces. In order to know how the temperature of the tread samples changes during the tests, an optical pyrometer pointing at the tread sample, measures the temperature on the contact surface in the initial position; while a thermocouple introduced into the tread block to record the temperature data during the test. In [52] are discussed the friction coefficient measurements for six tyre materials done on concrete and asphalt at 18 °C and 27 °C. Also this test bench offers the possibility to operate in both dry and wet conditions.

At Aalto University of Technology (Helsinki Finland), another example of this kind of LFT, called mini-μ-road (MMR) FT, was developed [56,57]. The MMR, presented in Figure 15, was specifically designed for low-friction testing [58].

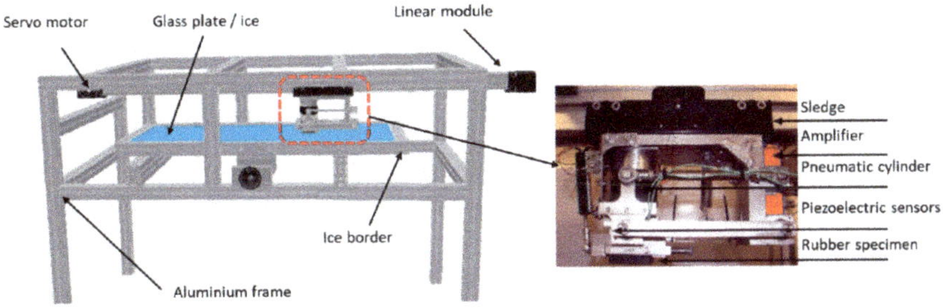

Figure 15. Representation of Mini-mu-road linear friction tester (adapted from [58]).

The experimental device consists of an aluminium frame that contains both the interchangeable test surfaces and a linear motion unit. It is equipped with a servomotor, whit a power of 4 kW, to drive a sledge supported by a linear rail system. The sledge holds and pulls a rigid frame that incorporates both the tread block specimen and the test equipment. This frame allows to host: a pneumatic cylinder used to generate the normal load, a sample holder, two piezoelectric load cells which measure the contact force between tread sample and rough surface and an amplifier module. The entire system is located in a climate chamber and is controlled by LabView. The main technical specifications are summarized in Table 7.

Table 7. Technical specification of mini-mu-road linear friction tester.

	Min	Max	
Speed	0.01	1.1	(m/s)
Normal force	-	1200	(N)
Temperature	−14	+50	(°C)
Sliding distance	-	1000	(mm)
Tread block sample	60 × 60 × 10		(mm^3)

In their paper [59] Kärkimaa, and Tuononen presented the results of a study conducted on tread rubber sample sliding on an asphalt road specimen. The tyre tread samples were provided by Nokian Tyres. The tests were conducted under different sliding conditions, varying the velocity and temperature. Measurements for different velocities from 1 mm/s to 1000 mm/s have been carried out at different temperatures with uncertainty about +/− 1 °C.

This device has a layout very similar to the one at Vienna University, the only technical difference is the maximum sliding length that the device can reach, which in this case is about 1000 mm. Among the examined devices, this is the second for sliding length.

Le Gal [60] in his Ph.D. thesis worked on a characterization of the friction coefficient by means of two different testing methods. Previously, a stationary friction experiment was performed using a modified Zwick universal test rig and subsequently a modified version of the MTS biaxial servo-hydraulic testing facility was used to extend the range of measurements and to simulate the typical loads in tyre application. Figure 16 shows a complete view of the MTS modified test facility layout.

It consists of a vertical cylinder positioned at the base of machine's floor, connected in series with a biaxial load cell for measures both the normal and the tangential force. An aluminium tank mounted on the load cell holds an interchangeable test surface and allows the possibility to use a liquid in order to simulate wet condition. A second cylinder, perpendicular to the vertical axis, on which is mounted an aluminium plate allows both the horizontal movement, and the possibility to fix the rubber sample with a maximum size of 80 × 80 mm^2; both axes (cylinder) are displacement controlled. The contact pressure between the rubber sample and rough surface can be varied assigning different displacement values to the vertical axis; on the other side, assigning different displacement time histories to the horizontal axis, it is possible to obtain different sliding speeds. The main technical specifications are summarized in Table 8.

(a)

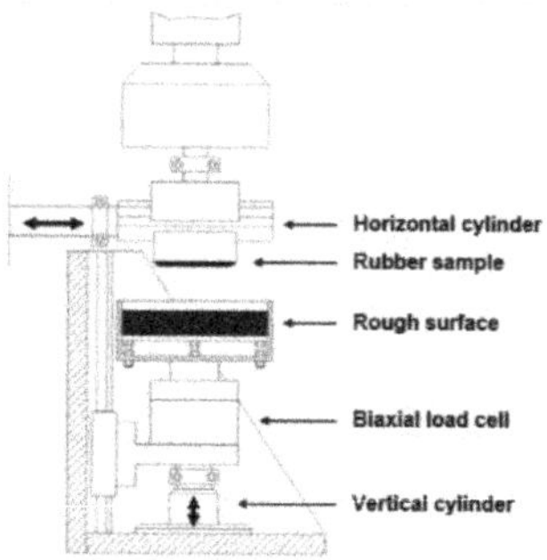

(b)

Figure 16. Picture of a modified biaxial MTS testing machine for friction tests (**a**) and schematic representation (**b**) [60].

Table 8. Technical specification of the modified biaxial MTS testing machine for friction tests.

	Min	Max	
Speed	0.1	0.4	(m/s)
Normal force	-	25	(kN)
Temperature	-	Amb	(°C)
Sliding distance	-	50	(mm)
Tread block sample	50 × 50 × 2		(mm^3)

Figure 17 depicts a typical measure of the contact forces. Figure refers to tests conducted at the sliding speed of 4 mm/s on a styrene-butadiene rubber (SBR) sample filled with 60 phr N339 and at the experimental pressure of 0.25 MPa between the rubber sample and the fine asphalt sample.

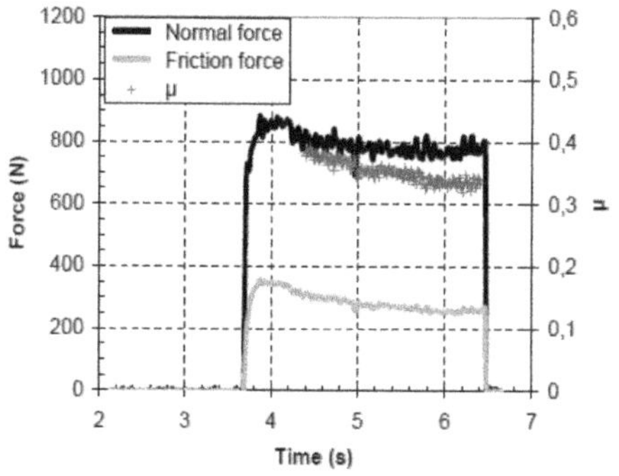

Figure 17. Measured normal and friction forces at constant velocity of 4 mm/s [60].

The high-speed tribometer developed at the Institute of Dynamics and Vibration Research (IDS) at the Leibniz Universität Hannover [61], also belongs to the LFT rubber sample in motion category. Such tester, also called HiLiTe, along with HSLFT manufactured by Altracon [62,63] is representative for a class of test rigs able to reach high velocities. The HiLiTe test machine, shown in Figure 18, was designed and developed in such a way to be able to simulate all relevant tyre testing conditions.

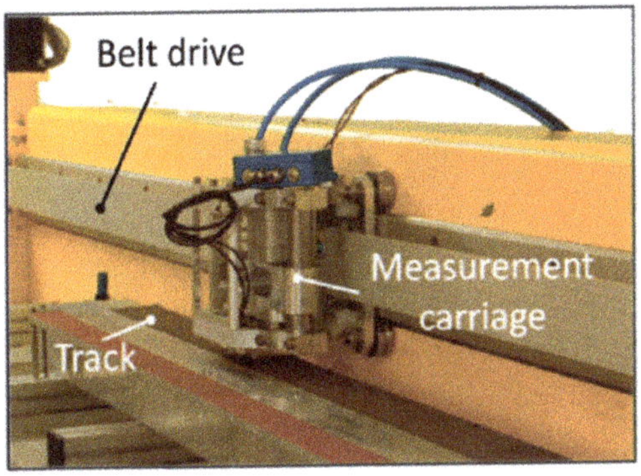

Figure 18. HiLiTe test rig [64].

It consists of a 5 m long linear guide rail in which a carriage is driven by a servomotor with a toothed belt and performs a linear movement. The carriage mounts all the test equipment and the rubber sample. Embedded on the carriage is a pre-stressed helical spring by means of which it is possible to generate the normal load between the sample and the road surface. It is actuated by a pneumatic actuator through which it is possible to set the normal force in the range of 23–1000 N. The carriage also contains a bi-axial piezoelectric force transducer which measures the normal and tangential components of the contact force between the tread sample and the test surface. The track can be equipped with any surface. The entire test bench is located in a climatic chamber so that experimental investigations can be conducted in a temperature range from −25 °C to 60 °C. The climate chamber also allows humidity control. The main technical specifications are summarized in Table 9.

Table 9. Technical specification of HiLiTe LFT.

	Min	Max	
Speed	-	10.0	(m/s)
Normal force	23	1000	(N)
Temperature	−25	60	(°C)
Sliding distance	-	5000	(mm)
Tread block sample	80 × 20		(mm^2)

The HiLiTe machine allows to use a number of different test tracks to reproduce a great variety of outdoor environmental conditions.

The Researchers of University of Hannover have performed several experimental investigations by means of HiLiTe [65,66] adopting concrete and asphalt tracks for dry and wet testing as well as test tracks made from ice and snow. In their paper Rosu et al. [64] performed an experimental investigation of the contact between an aircraft tyre rubber and rough surface. The experimental results, shown in Figure 19, were conducted by a rubber tread sample, measuring 20 × 20 × 10 mm^3, on an asphalt test track. Measurements at fixed sliding speed of 4 m/s with varying the load, by step, at four different temperatures have been carried out.

This is the only test rig, among those discussed in this review, with a top speed reaching 10 m/s, and the only device with a five-meter-long test track. Its technical specifications allow to explore the widest range conditions of contact between tyre and road.

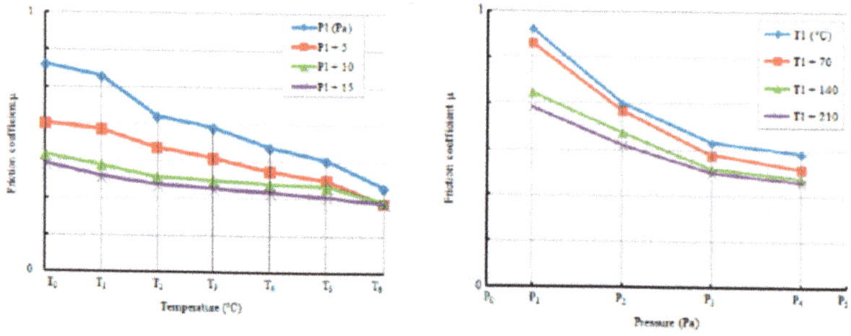

(a) (b)

Figure 19. Friction coefficient measurements, as function of the temperature (**a**) as function of contact pressure (**b**) [64].

4.3. On Field Friction Tester

It was decided to group this type of FT into a separate category, although the type of movement of the sample is still linear since these FTs have been specially designed to allow measurement "on field".

The FT designed and developed at the Centre for Tire Research (CenTiRe), a consortium of tyre and tyre-related industry members with two word-class universities (Virginia Tech and the University of Akron), belongs to this type of tester. This innovative device is an unmanned ground vehicle [67]. The robot shown in Figure 20, aims to reproduce all relevant tyre testing conditions on any surface and in both indoor and outdoor experiments.

(a) (b)

Figure 20. (**a**) The six-wheel small ground robot (**left**) [68]; (**b**) 3D representation of sample holder (adapted from [67]).

The device consists in a chassis with six wheels and an assembly of all the essential parts of the device, in which, four of these wheels are driven by a brushed permanent magnet DC motors, coupled directly to the wheels' shafts, and the other two are nondriven wheels. Two encoders are used to measure the rotational velocity of the wheels, one fixed to the driven wheel the other to the undriven wheel, so that the slip of the wheels can be calculated. Attached to the chassis a dragging arm holds and pulls the sample holder attached in series at load cell. The sample holder is built to fit a T-type sample with different size in the range of 25.4×25.4 mm^2 to 50.8×50.8 mm^2.

The main technical specifications are summarized in Table 10.

Table 10. Technical specification of the six-wheel small ground robot.

	Min	Max	
Speed	0.1	5.0	(m/s)
Normal force	-	140	(N)
Temperature	-	Amb	(°C)
Tread block sample	25.4 × 25.4	50.8 × 50.8	(mm^2)

The six-wheel small ground robot was developed to study friction and wear of a tread block sample on different surfaces and under different operating conditions. Emami et al. [67] showed the results of an experimental investigation conducted on a styrene-butadiene rubber (SBR) sample sliding on an asphalt track. The experimental results were conducted outdoor on a clean asphalt track at an outside temperature of 24 °C. The SBR sample used has dimension equal to 25.4 × 25.4 mm^2 and it is loaded with a weight of about 4.5 kg which corresponds to an equivalent pressure between the sample and the road of 81 kPa.

Although the robot was designed to perform measurements on field, it also allows indoor measurements, but these would require very large spaces specially if high speeds are to be achieved. The main limitation of this equipment consists of a particularly small normal load allowed.

Another type of on field FT was developed at the department of the "Forschungsgesellschaft Kraftfahrwesen mbH Aachen (fka) and the Institut für Kraftfahrzeuge (ika)" of RWTH Aachen University. The FT, called LiRep [69], is a portable device developed to study friction and wear of a tread block sample on real road surfaces under different operating conditions. The device, shown in Figure 21, consists of a four-wheeled frame to facilitate transport, therefore these wheels can be easily lifted or removed for performing the tests. The core of the device is a ball screw linear axis driven by a servo motor, that drags a "sensing head" (carriage/sledge) on which the tread-block sample, 3-axis force transducer and the loading weights are mounted. The "sensing-head" allows to install samples of a maximum size of 60 × 60 mm^2, and to add a calibrated weight up to 60 kg. Depending on the contact surface of the sample, these weights correspond to an equivalent pressure between the sample and the road up to 3.5 bar. All data acquisition system, cables and the control unit are located on the chassis. Furthermore, the test area can be covered with a climate chamber to study the effect of temperature on the friction coefficient. The main technical specifications are summarized in Table 11.

(a) (b)

Figure 21. (**a**) LiRep general view (left); (**b**) detail of two different road surface configurations [70].

Table 11. Technical specification of the LiRep device.

	Min	Max	
Speed	0.001	1.2	(m/s)
Normal force	-	60	(N)
Temperature	-	Amb	(°C)
Tread block sample	60 × 60		(mm^2)

Compared to other FT shown in this category, the main difference is that this device is not a mobile device but rather is attached to the floor, so in this case, it is easier to carry out laboratory tests since the speed and length of the sliding are not linked to the movement of the device.

5. Other Types

This section describes the FTs that not have a common contact mechanism or the same type of movement, for this reason they are classified as "others".

5.1. Pin on Disk Tribometers

The pin on disk (PoD) configuration is one of the most popular devices used to study the friction and wear. Introduced by the ASTM standards G133 [71] and G99 [72] for wear and erosion tests, both these devices are wide used also to study materials like tyre rubber.

Pin on disk (PoD) tribometers, in general, are designed and constructed to study the complex friction phenomena in many engineering applications such as railway wheels, automotive and aircraft brake systems, clutches, bearings, mechanical joints, tyre and biomedical materials. Conventional PoD machines provide a normal contact load between a stationary pin and a revolving disk and measure the resulting frictional force to evaluate the coefficient of friction [73–76]. PoD tests can use different contact geometries:

- point contact using a ball;
- flat contact, often performed using a cylinder;
- line contact performed adopting a cylinder aligned in a line contact, which reduces contact pressures a bit and increases the tested areas, but still represent a non-conformal contact and are easier to align than flat contacts.

An example of the above equipment is reported in the studies of Carbone et al. [77], in which the authors have investigated the friction properties of styrene-butadiene rubber (SBR) copolymer, furnished by Pirelli Tyre, by means a ball-on-disk configuration. The friction measurement has been carried out using the CSM Instruments Tribometer available at the Tribology Lab at the Department of Mechanical and Industrial Engineering (Politecnico di Bari, Bari, Italy). The device, shown in Figure 22, is composed by a rotating SBR disk in contact with a sphere made in polytetrafluoroethylene (PFTE) fixed to the holder and unable to rotate.

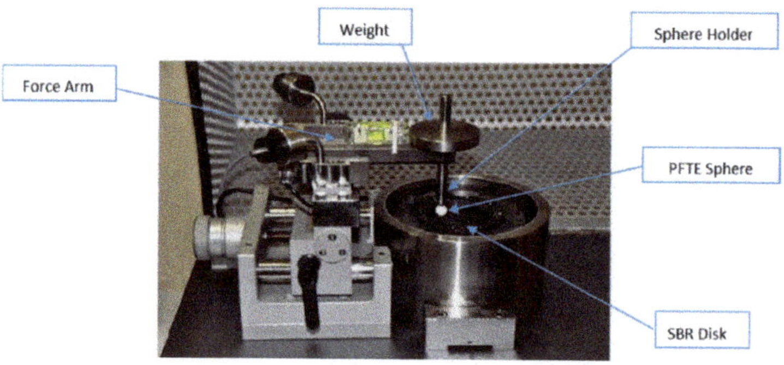

Figure 22. The CSM- instruments HT-Tribometer [77].

The sphere is mounted on a stiff lever, designed as a frictionless force transducer, and is loaded onto the test sample by applying calibrated weights so as to allow adjustment of the normal force in the range 0–10 N. The friction coefficient is determined during the test by measuring the deflection of the highly linear and precise elastic arm, with a resolution of 5 mN while wear coefficients for the ball and disk materials are calculated from the volume of material lost during the test. The instrument allows one to control the velocity of the disk and the radial position of the ball.

The main technical specifications are shown in a Table 12.

Table 12. Technical specification of the CSM instruments tribometer.

	Min	Max	
Speed	0.3	500	(rpm)
Normal force	-	10	(N)
Temperature	-	1000	(°C)
Test Disk Radius	30		(mm)

With the aim to investigate the viscoelastic contribution to friction, the authors chose a PFTE sphere to reduce friction and wear between the SBR disk and the sphere. The tests were conducted at the sliding velocity of 6 mm/s and for different normal loads. Figure 23 shows the comparison between experimental and numeric increase Δμ of the friction coefficient μ as the normal load is increased from 1 N to 5 N.

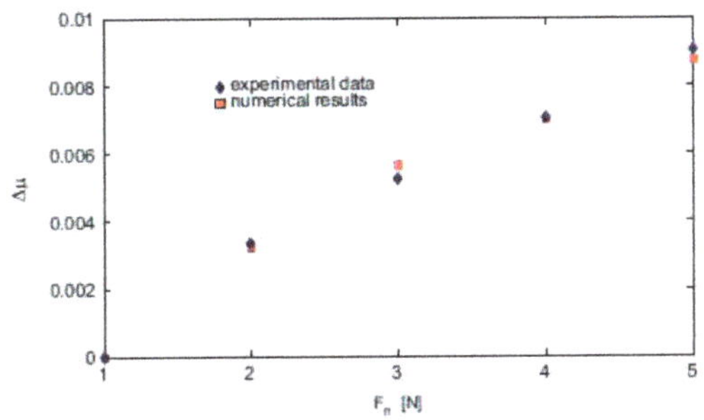

Figure 23. Comparison between experimental and numeric increase Δμ of the friction coefficient μ as the normal load is increased from 1 N to 5 N, blue rhomboids refer to experiments, red squared to numerical calculations [77].

This device is also designed to study friction and wear behaviour, since it allows one to vary the interfacial friction by replacing test spheres characterized by different roughness. The device allows measurement of friction in a small normal load range, and for this reason and also for the small contact surface, the device is not able to simulate the tyre/road contact conditions.

5.2. Dynamic Friction Tester

The Dynamic Friction Tester, shortly called DFT, was born like an on-field tester to characterize the paved surface frictional properties and shows a layout conceptually similar to the PoD tribometer. The DFT is equipped with a rotating disc with three connected rubber sample and does not have the fixed part (road surface) since this tester is usually used for friction measurements on the field, but in any case, as shown by a study conducted by Do et al. [78], the device allows laboratory measurements. This device designed and manufactured in Japan by the Nippo Sangyo Co., Ltd. has being used widely not only in Japan but also in the EU, the United States and many other countries by government agencies, construction companies, independent consulting companies, automotive manufacturers, tyre manufacturers, research institutions and universities among many other organizations [79]. In EU the leading user of this device is the "Institut Français des Sciences et Technologies des Transports, de l'Aménagement et des Réseaux (IFSTTAR)". Also, DFT has been chosen as the standard reference by IFI (International Friction Index) ASTM international Standards [80]. The DFT is reported in the ASTM G115 in the "vehicle pavement system" and the standard test procedure used in the United States can be found in the ASTM E-1911 standard [81]. The device, shown in Figure 24, looks like a box where the bottom side has three rubber sliders mounted on the lower surface of a disk that rotates with its plane parallel to the test surface. In the upper side there are the DC electric motor and the control unit.

(a) (b) (c)

Figure 24. Dynamic Friction Tester, (**a**) upper side, (**b**) bottom side, (**c**) rubber sliders [82].

The main technical specifications are shown in a Table 13.

Table 13. Technical specification.

	Min	Max	
Speed	-	25	(m/s)
Normal force	-	11.8	(N)
Temperature	-	Amb	(°C)
Test Disk Radius	167.5		(mm)

As mentioned, the DFT is commonly used for degeminating the road surface frictional properties, in several states in the world. A detailed analysis of the relevant aspects including the experimental results of pavement friction studies, using the DFT, is given by Rado et al. [83], and Kane et al. [82,84]. In the study conducted by Kane et al. [82] the results of the friction coefficient measurements conducted using the DFT in wet condition are presented, with the goal to develop and validate by means of the DFT a dynamic frictional contact model. During the measurement, the disk is accelerated to reach the target speed, after which, prior to reaching the desired speed, the water is applied and maintained during the entire measurement process through an irrigation system of the device. Once the set speed of the rubber pads is reached, the motor is turned off and the disk with the measuring pads is lowered in contact with the surface with a constant normal load. Each pad is loaded at 11.8 N, chosen due to the weight of the device and the rigidity of the pad holders. The speed of the pads decreases until it stops completely due to the friction generated between the pads and the contact surface, recorded during the deceleration phase from the set speed up to the stop.

The main advantages in the use of the DFT lies in the possibility to operate both indoor and outdoor, in the ease of use and in the possibility of making measurements in a wide range of speeds. The main drawback associated with the use of a commercial machine such as the DFT lays in the fact that it is not possible to study the friction coefficient of tyre tread compounds or to use tread blocks obtained from a tyre, because the friction pads are provided by the manufacturer. In addition, the DFT does not allow to vary the normal load on the sample.

5.3. British Pendulum Test Rig

The last device reported in this review is the British Pendulum (BP), widely accepted as a device for both field and laboratory friction testing. The BP test is described in ASTM E303 [85] as a laboratory testing method to find the skid resistance of pavement surface. It is a low-speed (<10 km/h) test which is related to surface micro-texture of road surface. A typical commercial device is shown in Figure 25, with its main components. In many researches, BP is used to perform the asphalt characterization [86,87], while this review is focused on the devices used to study the interactions between a tread block sample and the road surface.

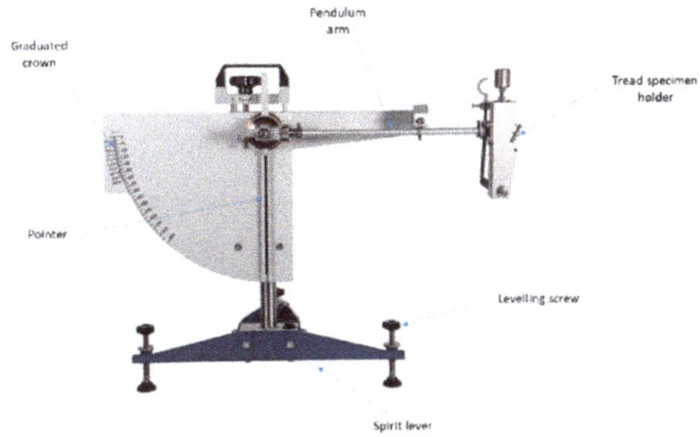

Figure 25. A typical commercial British Pendulum.

Ciaravola et al. [88], performed an experimental investigation of contact between tyre tread and rough surfaces by means an evolved version of a standard BP, developed starting from a BP at the Technical Centre Europe Bridgestone and customized at Department of Industrial Engineering, University of Naples Federico II (Naples, Italy). Later, Arricale et al. [89] performed an experimental investigation on tyre/road friction, between a tread block and real asphalt specimens by means an improved version of the British Pendulum developed by the UniNa Vehicle Dynamic Research Group, also called BP-EVO. The device showed in Figure 26, conserves the main components of the classic pendulum and is also equipped with a series of sensors allowing to further enrich the measurement dataset.

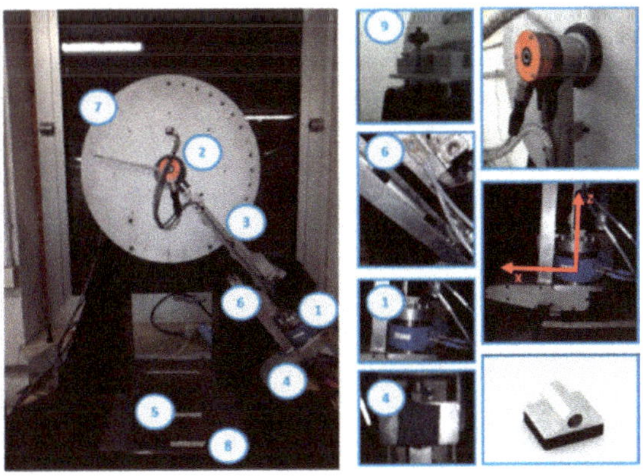

Figure 26. The British Pendulum Evo.

In detail the BP-EVO is composed by:

- an oscillating arm
- an encoder to measure the angular speed of the arm
- a tri-axial load cell, interposed between the rubber sample and the arm to measure the normal and tangential force
- a rubber specimen cut from a tyre tread
- a tank to host a road specimen

- a road specimen
- a pre-loading spring to vary the contact pressure at tyre/road interface
- a graduate crown located on the rigid frame of the device, in order to set the drop height, on which the sliding speed depends
- a levelling screw to set the height of the oscillating arm from the road specimen

The main technical specifications are showed in Table 14.

Table 14. Technical specification.

	Min	Max	
Speed	1	3.0	(m/s)
Normal force	-	250	(kPa)
Temperature	25	100	(°C)
Test Disk Sample	$25 \times 25 \times 5$		(mm^3)

The initial temperature of the tread sample can be varied up to 120 °C by means of an industrial heat gun and measured by an infrared pyrometer. The contact pressure can be varied using different springs. The BP, compared to the testers shown in this review, is certainly the easiest to use and, despite its simple layout, it offers the possibility to be employed in both dry and wet conditions adopting real asphalt and tread samples as a counter face. Since the pendulum is not driven by a motor, the sliding speed depends on the drop height; therefore, it is not possible making measurements in a wide range of sliding speeds. In addition, it should be noted that the range of sliding speeds may vary depending on the characteristics of the tread and the surface. For example, working with rough surfaces and tyres with high performance in terms of friction value, it is impossible to perform test at particularly low sliding speed. The reason is that, at low speed, the sample is stopped when come in contact with the surface, not allowing the friction measurement.

6. Summary and Conclusions

In this review, the studies and the developments carried out with the aim to test and analyse frictional and wear behaviour of viscoelastic materials, and of tyres tread in particular, were examined by comparing the different working principles of different so-called friction testers, with an in-depth focus on a selection of them, representative of each category.

In the authors' research activity on tyre analysis and friction modelling, the need of a wide overview of the state of the art in tribological testing on viscoelastic materials was experienced. Due to the lack of such kind of information, the aim of the review is to resume devices and methodologies developed in the far and recent past, in order to give suggestions and ideas to who could be interested to build a friction testing bench or to evaluate the optimal methodology to get data concerning interaction phenomena between a rigid and a viscoelastic material.

It has been found that the main variables affecting tyre friction in local contact are sliding speed, contact pressure, tread temperature, road (or in general counter-surface) roughness and compound viscoelastic characteristics. Consequently, most of the developed friction test benches allow to the user to vary such variables in a range due to space and to technological constraints.

Friction coefficient is expressed as the ratio between the tangential and the normal load acting on the tested specimen, and for this reason the rigs are usually equipped with load cells and with control systems that guarantee stability and precision in the test setup and in the conduction of the experiments.

A key obstacle encountered in the tyre friction testing regards the possibility to vary single working conditions, keeping the others constant. The difficulty is due to the mutual and deep interconnections among the phenomena arising at tyre/road contact, and in many of the described applications the solution has been fund involving very small variations in the working conditions. On the other hand, real tyre working conditions are characterized by significant values of speed and pressure, not easily reachable and controllable by laboratory benches. For this reason, "on field" friction testers have been developed, able to observe real contact conditions, despite the lower level of repeatability and precision of the measurements linked to the noise and the random nature of the asphalt profile.

The variation of contact pressure in the indoor benches is usually controlled by the actuation of controlled pneumatic systems, while the speed management is highly linked to the motion principle of the bench itself (linear, rotational, etc …).

Some of the illustrated devices, moreover, are installed in a climate chamber that allows a reliable control of temperature, highly affecting the tread polymers viscoelastic behaviour. Such solution improves the repeatability of the measurements and gives the opportunity to test the specimens also in uncommon or hardly reachable environmental conditions, but increases significantly the cost and the volume of the experimental setup.

Finally, some friction testers have been developed re-adapting systems conceived for different scopes (like pin on disk benches, often adopted for hardness and wear measurements in metal-metal sliding contact), substituting some elements with specimens of rubber to be tested.

As a final summary, a table of FT features (Table 15) has been developed, with the aim to provide a global overview of the described benches, each with its peculiar characteristics and advantages/disadvantages.

Table 15. Friction testers features resume.

Device	Sliding Speed (m/s)		Max Normal Force (N)	Temperature (°C)		Sliding Distance (m)	Tread Block Sample (mm²)	Operating Condition
	Min	*Max*		*Min*	*Max*			*Wet/ lubrication*
Rolling test rig (Liu et al.)	-	6.0	-	Amb.		-	16 × 10	×
CID	0	17	300	Amb.		-	30 × 34	×
LFT (Lorenz et al.)	5×10^{-6}	1×10^{-3}	26	−10	+120	-	20 × 20	×
LFT (A. Lang and M. Kluppel)	1×10^{-4}	1.0	280	+2	+100	-	20 × 20	✓
LFT (O'Neil et al.)	1×10^{-6}	0.05	-	−40	+180	0.07	35 × 40	×
LFT (O. Lahayne et al.)	1×10^{-5}	1.1	4800	−30	+40	0.3	80 × 20	✓
LFT, mini-μ-road (Aalto University)	1×10^{-5}	1.1	1200	−14	+50	1.0	60 × 60	✓
LFT (A. Le Gal)	1×10^{-4}	0.4	25,000	Amb.		0.05	50 × 50	✓
LFT, HiLiTe (Leibniz Universität Hannover)	-	10	1000	−25	+60	5.0	80 × 20	✓
Six-wheel small ground robot	0.1	5.0	140	Amb.		-	25.4 × 25.4 and 50.8 × 50.8	✓
LiReP	1×10^{-3}	1.2	60	Amb.		-	60 × 60	✓
CSM Instruments Tribometer	8×10^{-4}	1.3	10	-	+1000	-	-	✓
DFT	-	25	11.8	Amb.		-	20 × 16	✓
BP-EVO	1.0	3.0	160	Amb.			25 × 25	✓

In conclusion, friction, and in particular friction between a viscoelastic material and a randomly rough and rigid surface, still represents a field in which significant efforts in terms of research and testing are daily spent. Each of the benches reported and analysed is useful to study some specific aspect of very complex phenomena, but it seems that an ultimate technological solution, able to reproduce the global effects involved in tyre/road sliding contact in a fully satisfying way, has not been developed yet. The improvements in control systems and in the accuracy of the measurement devices are leading to increasingly evolved benches, able to lead researchers and industries, interested in a fundamental topic for mobility, road safety and polymers science, towards a deeper understanding of contact mechanics.

Author Contributions: Conceptualization, A.G. and F.F.; methodology, A.G. and G.A.D.; validation, A.S. and F.F.; formal analysis, A.S. and F.F.; investigation, A.G. and G.A.D.; data curation, G.A.D.; writing—original draft preparation, G.A.D., A.G., F.F. and A.S.; writing—review and editing, A.G., F.F. and A.S.; visualization, G.A.D. and F.F.; supervision, A.G. and A.S. All authors have read and agreed to the published version of the manuscript.

Funding: This research received no external funding.

Conflicts of Interest: The authors declare no conflict of interest.

References

1. Persson, B.N.J. *Sliding Friction: Physical Principles and Applications*; Springer Science & Business Media: Berlin/Heidelberg, Germany, 2013; ISBN 3662042835.
2. Dowson, D. *History of Tribology*; Longman: London, UK, 1979.
3. Hutchings, I.M. Leonardo da Vinci's studies of friction. *Wear* **2016**, *360–361*, 51–66.
4. Gnecco, E.; Meyer, E.; Gnecco, E.; Meyer, E. *Elements of Friction Theory and Nanotribology*; Cambridge University Press: Cambrige, UK, 2015.
5. Popov, V.L. *Contact Mechanics and Friction: Physical Principles and Applications*; Springer: Berlin/Heidelberg, Germany, 2017; ISBN 9783662530818.
6. Olsson, H.; Åström, K.J.; Canudas De Wit, C.; Gäfvert, M.; Lischinsky, P. Friction Models and Friction Compensation. *Eur. J. Control* **1998**, *4*, 176–195.
7. Angerhausen, J.; Murrenhoff, H.; Dorogin, L.; Scaraggi, M.; Lorenz, B.; Persson, B.N.J. Influence of anisotropic surfaces on the friction behaviour of hydraulic seals. In Proceedings of the BATH/ASME 2016 Symposium on Fluid Power and Motion Control American Society of Mechanical Engineers Digital Collection, Bath, UK, 7–9 September 2016.
8. Li, K.; Ergonomics, C.C.-A. *Undefined The Effect of Shoe Soling Tread Groove width on the Coefficient of Friction with Different Sole Materials, Floors, and Contaminants*; Elsevier: Amsterdam, The Netherlands, 2004.
9. Heidt, R.S.; Dormer, S.G.; Cawley, P.W.; Scranton, P.E.; Losse, G.; Howard, M. Differences in friction and torsional resistance in athletic shoe-turf surface interfaces. *Am. J. Sports Med.* **1996**, *24*, 834–842. [PubMed]
10. Wriggers, P.; Reinelt, J. Multi-scale approach for frictional contact of elastomers on rough rigid surfaces. *Comput. Methods Appl. Mech. Eng.* **2009**, *198*, 1996–2008.
11. Brown, R. Friction and Abrasion. In *Physical Test Methods for Elastomers*; Springer International Publishing: Berlin/Heidelberg, Germany, 2018; pp. 267–289.
12. G115-10 A. *G115-10 A Standard Guide for Measuring and Reporting Friction Coefficients*; ASTM International: West Conshohocken, PE, USA, 2004.
13. ISO. *ISO-15113. Rubber Determination of Frictional Properties*; ISO: Oxford, UK, 2005.
14. Bowden, F.; Solid, D.T. *The Adhesion Of Solids In The Structure an Properties of Solid Surface*; ACS Publications: Washington, DC, USA, 1953.
15. Bowden, F.; Tabor, D. *The Friction and Lubrication of Solids*; Clarendon Press: Oxford, UK, 2001.
16. Kummer, H.W. Unified Theory of Rubber and Tire Friction. Pennsylvania State University: University Park, PA, USA, 1966.
17. Persson, B.N.J.; Albohr, O.; Tartaglino, U.; Volokitin, A.I.; Tosatti, E. On the nature of surface roughness with application to contact mechanics, sealing, rubber friction and adhesion. *J. Phys. Condens. Matter* **2005**, *17*, R1–R62. [CrossRef]
18. Persson, B.N.J. Rubber friction: Role of the flash temperature. *J. Phys. Condens. Matter* **2006**, *18*, 7789–7823.
19. Grosch, K.A. Rubber friction and its relation to tire traction. *Rubber Chem. Technol.* **2007**, *80*, 379–411.
20. Moldenhauer, P. *Modellierung und Simulation der Dynamik und des Kontakts von Reifenprofilblöcken*; University of Freiberg: Freiberg, Germany, 2010.
21. Liang, H.; George, T.E.; Glenn, W.M. Lubrication and Tribology Fundamentals. *Fuels Lubr. Handb. Technol. Prop. Perform. Test. ASTM Int.* **2003**, *19*, 525–558.
22. Hutchings, I.; Gee, M.; Santner, E. Friction and Wear. In *Springer Handbook of Metrology and Testing*; Springer: Berlin/Heidelberg, Germany, 2011; pp. 743–768.
23. Moore, D.F. *The Friction And Lubrication of Elastomers*; Pergamon: Oxford, UK, 1972.
24. Moore, D.F. *The Friction of Pneumatic Tires*; Elsevire: Amsterdam, The Netherlands, 1975.
25. Grosch, K.A. The relation between the friction and visco-elastic properties of rubber. *Proc. R. Soc. Lond. Ser. A Math. Phys. Sci.* **1963**, *274*, 21–39.
26. Roberts, A.D. A guide to estimating the friction of rubber. *Rubber Chem. Technol.* **1992**, *65*, 673–686.

27. Genovese, A.; Carputo, F.; Ciavarella, M.; Farroni, F.; Papangelo, A.; Sakhnevych, A. Analysis of multiscale theories for viscoelastic rubber friction. In Proceedings of the Lecture Notes in Mechanical Engineering, Rome, Italy, 15–19 September 2020; pp. 1125–1135.
28. Greenwood, J.A.; Williamson, J.B.P. Contact of nominally flat surfaces. *Proc. R. Soc. Lond. Ser. A Math. Phys. Sci.* **1966**, *295*, 300–319.
29. Archard, J. Elastic deformation and the laws of friction. *Proc. R. Soc. London. Ser. A. Math. Phys. Sci.* **1957**, *243*, 190–205.
30. Bush, A.W.; Gibson, R.D.; Thomas, T.R. The elastic contact of a rough surface. *Wear* **1975**, *35*, 87–111. [CrossRef]
31. Bush, A.W.; Gibson, R.D.; Keogh, G.P. The limit of elastic deformation in the contact of rough surfaces. *Mech. Res. Commun.* **1976**, *3*, 169–174. [CrossRef]
32. Heinrich, G. Hysteresis friction of sliding rubbers on rough and fractal surfaces. *Rubber Chem. Technol.* **1997**, *70*, 14. [CrossRef]
33. Klüppel, M.; Heinrich, G. Rubber friction on self-affine road tracks. *Rubber Chem. Technol.* **2000**, *73*, 578–606. [CrossRef]
34. Le Gal, A.; Klüppel, M. Investigation and modelling of rubber stationary friction on rough surfaces. *J. Phys. Condens. Matter* **2008**, *20*, 015007. [CrossRef]
35. Persson, B.N.J. Theory of rubber friction and contact mechanics. *J. Chem. Phys.* **2001**, *115*, 3840–3861. [CrossRef]
36. Carbone, G.; Bottiglione, F. Asperity contact theories: Do they predict linearity between contact area and load. *J. Mech. Phys. Solids* **2008**, *56*, 2555–2572. [CrossRef]
37. Persson, B.N.J. On the elastic energy and stress correlation in the contact between elastic solids with randomly rough surfaces. *J. Phys. Condens. Matter* **2008**, *20*, 312001. [CrossRef]
38. Carbone, G.; Bottiglione, F. Contact mechanics of rough surfaces: A comparison between theories. *Meccanica* **2011**, *46*, 557–565. [CrossRef]
39. Genovese, A.; Farroni, F.; Papangelo, A.; Ciavarella, M. A Discussion on Present Theories of Rubber Friction, with Particular Reference to Different Possible Choices of Arbitrary Roughness Cutoff Parameters. *Lubricants* **2019**, *7*, 85.
40. Liu, F.; Sutcliffe, M.P.F.; Graham, W.R. Modeling of tread block contact mechanics using linear viscoelastic theory. *Tire Sci. Technol.* **2008**, *36*, 211–226. [CrossRef]
41. Liu, F.; Sutcliffe, M.P.F.; Graham, W.R. Prediction of tread block forces for a free-rolling tyre in contact with a smooth road. *Wear* **2010**, *269*, 672–683.
42. Lundberg, O.E.; Kari, L.; Arteaga, I.L. A compact internal drum test rig for measurements of rolling contact forces between a single tread block and a substrate. *Measurement* **2017**, *103*, 370–378. [CrossRef]
43. Lorenz, B.; Oh, Y.R.; Nam, S.K.; Jeon, S.H.; Persson, B.N.J. Rubber friction on road surfaces: Experiment and theory for low sliding speeds. *J. Chem. Phys.* **2015**, *142*, 194701. [CrossRef]
44. Kawasaki, S.; Tada, T.; Persson, B.N.J. Adhesion and friction between glass and rubber in the dry state and in water: Role of contact hydrophobicity. *Soft Matter* **2018**, *14*, 5428–5441. [CrossRef]
45. Tolpekina, T.V.; Persson, B.N.J. Adhesion and Friction for Three Tire Tread Compounds. *Lubricants* **2019**, *7*, 20.
46. Lorenz, B.; Persson, B.N.J.; Dieluweit, S.; Tada, T. Rubber friction: Comparison of theory with experiment. *Eur. Phys. J. E* **2011**, *34*, 129.
47. Lorenz, B.; Persson, B.N.J.; Fortunato, G.; Giustiniano, M.; Baldoni, F. Rubber friction for tire tread compound on road surfaces. *J. Phys. Condens. Matter* **2013**, *25*, 95007.
48. Williams, M.L.; Landel, R.F.; Ferry, J.D. The Temperature Dependence of Relaxation Mechanisms in Amorphous Polymers and Other Glass-forming Liquids. *J. Am. Chem. Soc.* **1955**, *77*, 3701–3707. [CrossRef]
49. Lang, A.; Klüppel, M. Influences of temperature and load on the dry friction behaviour of tire tread compounds in contact with rough granite. *Wear* **2017**, *380–381*, 15–25. [CrossRef]
50. O'Neill, A.; Gruber, P.; Watts, J.F.; Prins, J. Predicting Tyre Behaviour on Different Road Surfaces. In *The AVSD Symposium on Dynamics of Vehicles on Roads and Tracks*; CRC Press: Boca Raton, FL, USA, 2019.
51. Gruber, P.; Fina, E. Tyres and Roads: Predicting Friction. *Veh. Ride Handl. Spec. Eng. Improv. Exp.* **2017**. Available online: https://epubs.surrey.ac.uk/844667/1/Tyres%20and%20Roads.pdf (accessed on 16 September 2020).

52. Lahayne, O.; Eberhardsteiner, J.; Reihsner, R. Tribological Investigations carried out by a linear friction tester. *Trans. Famena* **2009**, *33*, 15–22.
53. Lahayne, O.; Eberhardsteiner, J. *Investigation of the Temperature Behaviour of Sliding Rubber Materials*; WIT Transaction on Engineering Sciences WIT press: Ashurst, Southampton, UK, 2007; Volume 55, p. 155.
54. Schwaiger, H. *Entwicklung einer Prüfeinrichtung zur Untersuchung des Traktionsverhaltens von Gummiproben auf Verschiedenen Oberflächen*; Vienna University of Technology: Vienna, Austria, 1996.
55. Eberhardsteiner, J.; Fidi, W.; Lahayne, O.; Ag, C. Experimental Investigation of Adhesive Friction Properties of Rubber on Wet Surfaces. In Proceedings of the 5th Danubia-Adria Symposium on Experimental Methods in Solid Mechanics, Bertinoro, Italy, 30 September–3 October 1998.
56. Vainikka, J. *Development Of Rubber Friction Measuring Device*; Helsinki University of Technology: Espoo, Finland, 2001.
57. Vainikka, J.; Pirjola, H.; Sainio, P.; Juhala, M. Mini-mu-Road and Lontra-Arctic Friction Measuring Devices in Laboratory and on Field and their Correlation. In Proceedings of the FISITA World Automotive Congress, Barcelona, Spain, 23–27 May 2004.
58. Rantonen, M.; Tuononen, A.J.; Sainio, P. Measuring stud and rubber friction on ice under laboratory conditions. *Int. J. Veh. Syst. Model. Test.* **2012**, *7*, 194–207. [CrossRef]
59. Kärkimaa, J.; Tuononen, A.J. Experimental study on velocity and temperature dependency of rubber-asphalt friction. In *Proceedings of the The Dynamics of Vehicles on Roads and Tracks. In Proceedings of the 24th Symposium of the International Association for Vehicle System Dynamics (IAVSD 2015), Graz, Austria, 17–21 August 2015*; CRC Press: Boca Raton, FL, USA, 2015; p. 453.
60. Le Gal, A. Investigation and Modelling of Rubber Stationary Friction on Rough Surfaces. Ph.D. Thesis, University of Hannover, Hannover, Germany, 2007.
61. Moldenhauer, P.; Ripka, S.; Gäbel, G.; Kröger, M. Tire tread block dynamics: Experimental and numerical investigating of sliding friction. *Tire Technol. Int. Annu. Rev.* **2008**, *123*, 95–100.
62. Altracon Linear Friction Tester. Available online: http://www.altracon.com/HighSpeedLinearFrictionTester.php (accessed on 17 July 2020).
63. Ralf, B. Dream machines? HSLFT High speed linear friction tester. Tire Technology International. July 2011; 22–25.
64. Rosu, I.; Elias-Birembaux, H.L.; Lebon, F.; Lind, H.; Wangenheim, M. Experimental and numerical simulation of the dynamic frictional contact between an aircraft tire rubber and a rough surface. *Lubricants* **2016**, *4*, 29. [CrossRef]
65. Ripka, S.; Gäbel, G.; Wangenheim, M. Dynamics of a Siped Tire Tread Block—Experiment and Simulation. *Tire Sci. Technol.* **2009**, *37*, 323–339. [CrossRef]
66. Ripka, S.; Wangenheim, M. Simulation of a siped tire tread block by an Euler-Bernoulli beam. In *Proceedings of the PAMM: Proceedings in Applied Mathematics and Mechanics*; Wiley Online Library: Hoboken, NJ, USA, 2009; Volume 9, pp. 531–532.
67. Emami, A.; Khaleghian, S.; Bezek, T.; Taheri, S. Design and development of a new portable test setup to study friction and wear. *Proc. Inst. Mech. Eng. Part J J. Eng. Tribol.* **2020**, *234*, 730–742. [CrossRef]
68. Khaleghian, S.; Taheri, S. Terrain classification using intelligent tire. *J. Terramech.* **2017**, *71*, 15–24. [CrossRef]
69. Hüsemann, T.; Bachmann, C.; Winter, S.; Henrichmöller, D. Mobile testing equipment for the measurement of tire/road friction coefficients. *ATZ Worldw. eMagazine* **2011**, *113*, 56–60. [CrossRef]
70. Hartung, F.; Kienle, R.; Götz, T.; Winkler, T.; Ressel, W.; Eckstein, L.; Kaliske, M. Numerical determination of hysteresis friction on different length scales and comparison to experiments. *Tribol. Int.* **2018**, *127*, 165–176. [CrossRef]
71. ASTM-G133-05. *Standard Test Method For Measuring Paved Surface Frictional Properties Using The Dynamic Friction Tester*; ASTM International: West Conshohocken, PE, USA, 2010.
72. ASTM-G99. Standard Test Method for Wear Testing with a Pin-on-Disk Apparatus. In *Annual Book of ASTM Standards*; ASTM International: West Conshohocken, PA, USA, 1999; p. 401.
73. Yoon, E.-S.; Kong, H.; Kwon, O.-K.; Oh, J.-E. Evaluation of frictional characteristics for a pin-on-disk apparatus with different dynamic parameters. *Wear* **1997**, *203*, 341–349. [CrossRef]
74. Abdo, J.; Zaier, R. A novel pin-on-disk machine for stick-slip measurements. *Mater. Manuf. Process.* **2012**, *27*, 751–755. [CrossRef]

75. Saikko, V. A multidirectional motion pin-on-disk wear test method for prosthetic joint materials. *J. Biomed. Mater. Res. Off. J. Soc. Biomater. Jpn. Soc. Biomater. Aust. Soc. Biomater.* **1998**, *41*, 58–64. [CrossRef]
76. Bezzazi, M.; Khamlichi, A.; Jabbouri, A.; Reis, P.; Davim, J.P. Experimental characterization of frictional behaviour of clutch facings using Pin-on-disk machine. *Mater. Des.* **2007**, *28*, 2148–2153. [CrossRef]
77. Carbone, G.; Putignano, C. A novel methodology to predict sliding and rolling friction of viscoelastic materials: Theory and experiments. *J. Mech. Phys. Solids* **2013**, *61*, 1822–1834. [CrossRef]
78. Do, M.-T.; Cerezo, V.; Beautru, Y.; Kane, M. Modeling of the connection road surface microtexture/water depth/friction. *Wear* **2013**, *302*, 1426–1435. [CrossRef]
79. Nippo Sangyo Co., Ltd. Available online: http://www.nippou.com/en/products/dft.html (accessed on 17 July 2020).
80. Henry, J.J.; Abe, H.; Kameyama, S.; Tamai, A.; Kasahara, A.; Saito, K. Determination of the international friction index (IFI) using the Circular Texture Meter (CTM) and the Dynamic Friction Tester (DFT). In Proceedings of the SURF 2000: Fourth International Symposium on Pavement Surface Characteristics on Roads and AirfieldsWorld Road Association-PIARC, Nantes, France, 22–24 May 2000.
81. ASTM-E1911. *Standard test Method for Measuring Paved Surface Frictional Properties Using the Dynamic Friction Tester*; ASTM International: West Conshohocken, PE, USA, 2009.
82. Kane, M.; Cerezo, V. A contribution to tire/road friction modeling: From a simplified dynamic frictional contact model to a "Dynamic Friction Tester" model. *Wear* **2015**, *342*, 163–171. [CrossRef]
83. Rado, Z.; Kane, M. An initial attempt to develop an empirical relation between texture and pavement friction using the HHT approach. *Wear* **2014**, *309*, 233–246. [CrossRef]
84. Kane, M.; Do, M.-T.; Cerezo, V.; Rado, Z.; Khelifi, C. Contribution to pavement friction modelling: An introduction of the wetting effect. *Int. J. Pavement Eng.* **2019**, *20*, 965–976. [CrossRef]
85. ASTM-E303. *Standard Test Method for Measuring Surface Frictional Properties Using the British pendulum Tester*; ASTM International: West Conshohocken, PE, USA, 2013.
86. Qasrawi, H.; Asi, I. Effect of bitumen grade on hot asphalt mixes properties prepared using recycled coarse concrete aggregate. *Constr. Build. Mater.* **2016**, *121*, 18–24. [CrossRef]
87. Ling, T.-C.; Nor, H.M.; Lim, S.-K. Using recycled waste tyres in concrete paving blocks. In *Proceedings of Institution of Civil Engineers—Waste and Resource Management*; Thomas Telford Ltd.: London, UK, 2010; Volume 163, pp. 37–45. [CrossRef]
88. Ciaravola, V.; Farroni, F.; Fortunato, G.; Russo, M.; Russo, R.; Sakhnevych, A.; Timpone, F. An Evolved Version of the British Pendulum Tester for the Experimental Investigation of Contact Between Tire Tread and Rough Surfaces. *Eeng. Lett.* **2017**, *25*, 1.
89. Arricale, V.M.; Carputo, F.; Farroni, F.; Sakhnevych, A.; Timpone, F. Experimental Investigations on Tire/Road Friction Dependence from Thermal Conditions Carried out with Real Tread Compounds in Sliding Contact with Asphalt Specimens. *Trans Tech Publ.* **2019**, *813*, 261–266. [CrossRef]

MDPI
St. Alban-Anlage 66
4052 Basel
Switzerland
Tel. +41 61 683 77 34
Fax +41 61 302 89 18
www.mdpi.com

Lubricants Editorial Office
E-mail: lubricants@mdpi.com
www.mdpi.com/journal/lubricants